Volcanoes: Past and Present

by Edward Hull

PREFACE.

It has not been my object to present in the following pages even an approximately complete description of the volcanic and seismic phenomena of the globe; such an undertaking would involve an amount of labour which few would be bold enough to attempt; nor would it be compatible with the aims of the Contemporary Science Series.

I have rather chosen to illustrate the most recent conclusions regarding the phenomena and origin of volcanic action, by the selection of examples drawn from the districts where these phenomena have been most carefully observed and recorded under the light of modern geological science. I have also endeavoured to show, by illustrations carried back into later geological epochs, how the volcanic phenomena of the present day do not differ in kind, though they may in degree, from those of the past history of our globe. For not only do the modes of eruption of volcanic materials in past geological times resemble those of the present or human epoch, but the materials themselves are so similar in character that it is only in consequence of alterations in structure or composition which the original materials have undergone, since their extrusion, that any important distinctions can be recognised between the volcanic products of recent times and those of earlier periods.

I have, finally, endeavoured to find an answer to two interesting and important questions: (1) Are we now living in an epoch of extraordinary volcanic energy?--a question which such terrible outbursts as we have recently witnessed in Japan, the Malay Archipelago, and even in Italy, naturally suggest; and (2) What is the ultimate cause of volcanic action? On this latter point I am gratified to find that my conclusions are in accordance with those expounded by one who has been appropriately designated "the Nestor of Modern Geology," Professor Prestwich.

Within the last few years the study of the structure and composition of volcanic rocks, by means of the microscope brought to bear on their translucent sections, has added wonderfully to our knowledge of such rocks, and has become a special branch of petrological investigation. Commenced by Sorby, and carried on by Allport, Zirkel, Rosenbusch, Von Lasaulx, Teall, and many more enthusiastic students, it has thrown a flood of light upon our knowledge of the mutual relations of the component minerals of igneous masses, the alteration these minerals have undergone in some cases, and the conditions under which they have been erupted and consolidated. But nothing that has been observed has tended materially to alter conclusions arrived at by other processes of reasoning regarding volcanic phenomena, and for these we have to fall back upon observations conducted in the field on a more or less large scale, and carried on before, during, and after eruptions. Macroscopic and microscopic observations have to go hand in hand in the study of volcanic phenomena.

E. H.

CONTENTS.

PART I.

INTRODUCTION.

PAGE

Chap. I. Historic Notices of Volcanic Action

" II. Form, Structure, and Composition of Volcanic Mountains

" III. Lines and Groups of Active Volcanic Vents

" IV. Mid-ocean Volcanic Islands

PART II.

EUROPEAN VOLCANOES.

Chap. I. Vesuvius

" II. Etna

" III. The Lipari Islands, Stromboli

" IV. The Santorin Group

" V. European Extinct or Dormant Volcanoes

" VI. Extinct Volcanoes of Central France

" VII. The Volcanic District of the Rhine Valley

PART III.

DORMANT OR MORIBUND VOLCANOES OF OTHER PARTS OF THE WORLD.

Chap. I. Dormant Volcanoes of Palestine and Arabia

" II. The Volcanic Regions of North America

" III. Volcanoes of New Zealand 1

PART IV.

TERTIARY VOLCANIC DISTRICTS OF THE BRITISH ISLES.

Chap. I. Antrim 154

" II. Succession of Volcanic Eruptions

" III. Island of Mull and Adjoining Coast

" IV. Isle of Skye

" V. The Scuir of Eigg

" VI. Isle of Staffa

PART V.

PRE-TERTIARY VOLCANIC ROCKS.

Chap. I. The Deccan Trap-series of India

" II. Abyssinian Table-lands

" III. Cape Colony

" IV. Volcanic Rocks of Past Geological Periods of the British Isles

PART VI.

SPECIAL VOLCANIC AND SEISMIC PHENOMENA.

Chap. I. The Eruption of Krakatoa in 1883

" II. Earthquakes

PART VII.

VOLCANIC AND SEISMIC PROBLEMS.

Chap. I. The Ultimate Cause of Volcanic Action

" II. Lunar Volcanoes

" III. Are we Living in an Epoch of Special Volcanic Activity?

APPENDIX.

A Brief Account of the Principal Varieties of Volcanic Rocks

PART I.

INTRODUCTION.

CHAPTER I.

HISTORIC NOTICES OF VOLCANIC ACTION.

There are no manifestations of the forces of Nature more calculated to inspire us with feelings of awe and admiration than volcanic eruptions preceded or accompanied, as they generally are, by earthquake shocks. Few agents have been so destructive in their effects; and to the real dangers which follow such terrestrial convulsions are to be added the feelings of uncertainty and revulsion which arise from the fact that the earth upon which we tread, and which we have been accustomed to regard as the emblem of stability, may become at any moment the agent of our destruction. It is, therefore, not surprising that the ancient Greeks, who, as well as the Romans, were close observers of the phenomena of Nature, should have investigated the causes of terrestrial disturbances, and should have come to some conclusions upon them in accordance with the light they possessed. These terrible forces presented to the Greeks, who clothed all the operations of Nature in poetic imagery and deified her forces, their poetical and mystical side; and as there was a deity for every natural force, so there was one for earthquakes and volcanoes. Vulcan, the deformed son of Juno (whose name bears so strange a resemblance to that of "the first artificer in iron" of the Bible, Tubal Cain), is condemned to pass his days under Mount Etna, fabricating the thunderbolts of Jove, and arms for the gods and great heroes of antiquity.

The Pythagoreans appear to have held the doctrine of a central fire (meson pyr) as the source of volcanic phenomena; and in the Dialogues of Plato allusion is made to a subterranean reservoir of lava, which, according to Simplicius, was in accordance with the doctrine of the Pythagoreans which Plato was recounting.[1] Thucydides clearly describes the effect of earthquakes upon coast-lines of the Grecian Archipelago, similar to that which took place in the case of the earthquake of Lisbon, the sea first retiring and afterwards inundating the shore. Pliny supposed that it was by earthquake avulsion that islands were naturally formed. Thus Sicily was torn from Italy, Cyprus from Syria, Euboea from Boeotia, and the rest; but this view was previously enunciated by Aristotle in his "Peri kosmou," where he states that earthquakes have torn to pieces many parts of the earth, while lands have been converted into sea, and that tracts once covered by the sea have been converted into dry land.

But the most philosophical views regarding terrestrial phenomena are those given by Ovid as having been held by Pythagoras (about B.C. 580). In the Metamorphoses his views regarding the interchange of land and sea, the effects of running water in eroding valleys, the growth of deltas, the effect of earthquakes in burying cities and diverting streams from their sources, are remarkable anticipations of doctrines now generally held.[2] But what most concerns us at present are his views regarding the changes which have come over volcanic mountains. In his day Vesuvius was dormant, but Etna was active; so his illustrations are drawn from the latter mountain; and in this connection he observes that volcanic vents shift their position. There was a time, he says, when Etna was not a burning mountain, and the time will come when it will cease to burn; whether it be that some caverns become closed up by the movements of the earth, or others opened, or whether the fuel is finally exhausted.[3] Strabo may be regarded as having originated the view, now generally held, that active volcanoes are safety-valves to the regions in which they are situated. Referring to the tradition recorded by Pliny, that Sicily was torn from Italy by an earthquake, he observes that the land near the sea in those parts was rarely shaken by earthquakes, since there are now orifices whereby fire and ignited matters and waters escape; but formerly, when the volcanoes of Etna, the Lipari Islands, Ischia, and others were closed up, the imprisoned fire and wind might have produced far more violent movements.[4]

The account of the first recorded eruption of Vesuvius has been graphically related by the younger Pliny in his two letters to Tacitus, to which I shall have occasion to refer further on.[5] These bring down the references to volcanic phenomena amongst ancient authors to the commencement of the Christian era; from all of which we may infer that the more enlightened philosophers of antiquity had a general idea that eruptions had their origin in a central fire within the interior of the earth, that volcanic mountains were liable to become dormant for long periods, and afterwards to break out into renewed activity, that there existed a connection between volcanic action and earthquakes, and that volcanoes are safety-valves for the regions around.

It is unnecessary that I should pursue the historical sketch further. Those who wish to know the views of writers of the Middle Ages will find them recorded by Sir Charles Lyell.[6] The long controversy carried on during the latter part of the eighteenth century between "Neptunists," led by Werner on the one side, and "Vulcanists," led by Hutton and Playfair on the other, regarding the origin of such rocks as granite and basalt, was finally brought to a close by the triumph of the "Vulcanists," who demonstrated that such rocks are the result of igneous fusion; and that in the cases of basalt and its congeners, they are being extruded from volcanic vents at the present day. The general principles for the classification of rocks as recognised in modern science may be regarded as having been finally established by James Hutton, of Edinburgh, in his Theory of the Earth,[7] while they were illustrated and defended by Professor Playfair in his work entitled, Illustrations of the Huttonian Theory of the Earth,[8] although other observers, such as Desmarest, Collini, and Guettard, had in other countries come to very clear views on this subject.

The following are some of the more important works on the phenomena of volcanoes and earthquakes published during the present century:--[9]

1. Poulett Scrope, F.R.S., Considerations on Volcanoes (1825). This work is dedicated to Lyell, his fellow-worker in the same department of science, and was undertaken, as he says, "in order to help to dispel that signal delusion as to the mode of action of the subtelluric forces with which the Elevation-Crater theory had mystified the geological world." The second edition was published in 1872.

2. This was followed by the admirable work, On the Extinct Volcanoes of Central France, published in 1826 (2nd edition, 1858), and is one of the most complete monographs on a special volcanic district ever written.

3. Dr. Samuel Hibbert, History of the Extinct Volcanoes of the Basin of Neuwied on the Lower Rhine (1832). Dr. Hibbert's work is one of remarkable merit, if we consider the time at which it was written. For not only does it give a clear and detailed account of the volcanic phenomena of the Eifel and the Lower Rhine, but it anticipates the principles upon which modern writers

account for the formation of river valleys and other physical features; and in working out the physical history of the Rhine valley below Mainz, and its connection with the extinct volcanoes which are found on both banks of that river, he has taken very much the same line of reasoning which was some years afterwards adopted by Sir A. Ramsay when dealing with the same subject. It does not appear that the latter writer was aware of Dr. Hibbert's treatise.

4. Leopold von Buch, Description Physique des Iles Canaries (1825), translated from the original by C. Boulanger (1836); Geognostische Reise (Berlin, 1809), 2 vols.; and Reise durch Italien (1809). From a large number of writings on volcanoes by this distinguished traveller, whom Alexander von Humboldt calls "dem geistreichen Forscher der Natur," the above are selected as being the most important. That on the Canaries is accompanied by a large atlas, in which the volcanoes of Teneriffe, Palma, and Lancerote, with some others, are elaborately represented, and are considered to bear out the author's views regarding the formation of volcanic cones by elevation or upheaval. The works dealing with the volcanic phenomena of Central and Southern Italy are also written with the object, in part at least, of illustrating and supporting the same theoretical views; with these we have to deal in the next chapter.

5. Dr. Charles Daubeny, F.R.S., Description of Active and Extinct Volcanoes, of Earthquakes, and of Thermal Springs, with remarks on the causes of these phenomena, the character of their respective products, and their influence on the past and present condition of the globe (2nd edition, 1848). In this work the author gives detailed descriptions of almost all the known volcanic districts of the globe, and defends what is called "the chemical theory of volcanic action"--a theory at one time held by Sir Humphrey Davy.

6. Wolfgang Sartorius von Waltershausen, Der 苦 na. This work possesses a melancholy interest from the fact that its distinguished author did not live to see its publication. Von Waltershausen, having spent several years in making an elaborate survey of Etna, produced an atlas containing numerous detailed maps, views, and drawings of this mountain and its surroundings, which were published at Weimar by Engelmann in 1858. A description in MS. to accompany the atlas was also prepared, but before it was printed, the author died, on the 16th October 1876. The MS. having been put into the hands of

the late Professor Arnold von Lasaulx by the publisher of the atlas, it was subsequently brought out under the care of this distinguished petrologist, who was so fully fitted for an undertaking of this kind.

7. Sir Charles Lyell in his Principles of Geology[10] devotes several chapters to the consideration of volcanic phenomena, in which, being in harmony with the views of his friend, Poulett Scrope, he combats the "elevation theory" of Von Buch, as applied to the formation of volcanic mountains, holding that they are built up of ashes, stones, and scori?blown out of the throat of the volcano and piled around the orifice in a conical form. Together with these materials are sheets of lava extruded in a molten condition from the sides or throat of the crater itself.

8. Professor J. W. Judd, F.R.S., in his able work entitled, Volcanoes: What they are, and what they teach,[11] has furnished the student of vulcanicity with a very complete manual of a general character on the subject. The author, having extensive personal acquaintance with the volcanoes of the south of Europe and the volcanic rocks of the British Isles, was well equipped for undertaking a work of the kind; and in it he supports the views of Lyell and Scrope regarding the mode of formation of volcanic mountains.

9. Sir Archibald Geikie, F.R.S., in his elaborate monograph[12] on the Tertiary Volcanic Rocks of the British Isles, has recorded his views regarding the origin and succession of the plateau basalts and associated rocks over the region extending from the north of Ireland to the Inner Hebrides; and in dealing with these districts in the following pages I have made extensive use of his observations and conclusions.

10. Report published by the Royal Society on the Eruption of Krakatoa--drawn up by several authors (1885)--and the work on the same subject by Chev. Verbeek, and published by the Government of the Netherlands (1886). In these works all the phenomena connected with the extraordinary eruptions of Krakatoa in 1883 are carefully noted and scientifically discussed, and illustrated by maps and drawings.

11. The Charleston Earthquake of August 31, 1886, by Captain Clarence Edward Dutton, U.S. Ordnance Corps. Ninth Annual Report of the United States Geological Survey, 1887-88, with maps and illustrations.

12. Amongst other works which may be consulted with advantage is that of Mr. T. Mellard Reade on The Origin of Mountain Ranges; the Rev. Osmond Fisher's Physics of the Earth; Professor G. H. Darwin and Mr. C. Davison on "The Internal Tension of the Earth's Crust," Philosophical Transactions of the Royal Society, vol. 178; Mr. R. Mallet, "On the Dynamics of Earthquakes," Trans. Roy. Irish Academy, vol. xxi.; Professor O'Reilly's "Catalogues of Earthquakes," Trans. Roy. Irish Academy, vol. xxviii. (1884 and 1888); and Mr. A. Ent. Gooch On the Causes of Volcanic Action (London, 1890). These and other authorities will be referred to in the text.

[1] See Julius Schwarez On the Failure of Geological Attempts made by the Greeks. (Edition 1888.)

[2] "Vidi ego, quod fuerat quondam solidissima tellus, Esse fretum. Vidi factas ex uore terras: Et procul ?pelago conch?jacuere marin? Et vetus inventa est in montibus anchora sumnis. Quodque fuit campus, vallem de cursus aquarum Fecit; et eluvie mons est deductus in uor: Eque paludosa siccis humus aret arenis; Quue sitim tulerant, stagnata paludibus hument. Hic fontes Natura novos emissit, at illuc Clausit: et antiquis concussa tremoribus orbis Fulmina prosiliunt...." --Lib. xv. 262.

[3] "Nec, qu?sulfureis ardet fornacibus, Etne Ignea semper erit; neque enim fuit ignea semper. Nam, sive est animal tellus, et vivit, habetque Spiramenta locis flammam exhalantia multis; Spirandi mutare vias, quotiesque movetur, Has finire potest, illas aperire cavernas: Sive leves imis venti cohibentur in antris; Saxaque cum saxis...." --Ibid., 340.

[4] Strabo, lib. vi.

[5] Tacitus, lib. vi. 16, 20.

[6] Principles of Geology, 11th edition, vol. i., ch. 3.

[7] 2 vols., Edin. (1795).

[8] Edin. (1802).

[9] A more extended list of early works will be found in Daubeny's Volcanoes (1848).

[10] 11th edition (1872).

[11] 4th edition (1888).

[12] "The History of Volcanic Action during the Tertiary Period in the British Isles," Trans. Roy. Soc., Edin. Vol. xxxv, (1888).

CHAPTER II.

FORM, STRUCTURE, AND COMPOSITION OF VOLCANIC MOUNTAINS.

The conical form of a volcanic mountain is so generally recognised, that many persons who have no intelligent acquaintance with geological phenomena are in the habit of attributing to all mountains having a conical form, and especially if accompanied by a truncated apex, a volcanic origin. Yet this is very far from being the fact, as some varieties of rock, such as quartzite, not unfrequently assume this shape. Of such we have an example in the case of Errigal, a quartzite mountain in Donegal, nearly 3000 feet high, which bears a very near approach in form to a perfect cone or pyramid, and yet is in no way connected, as regards its origin or structure, with volcanic phenomena. Another remarkable instance is that of Schehallion in Scotland, also composed of quartz-rock; and others may be found amongst the ranges of Islay and Jura, described by Sir A. Geikie.[1]

Notwithstanding, however, such exceptions, which might be greatly multiplied, the majority of cone-shaped mountains over the globe have a volcanic origin.[2] The origin of this form in each case is entirely distinct. In the case of quartzite mountains, the conical form is due to atmospheric influences acting on a rock of uniform composition, traversed by numerous joints and fissures crossing each other at obtuse angles, along which the rock breaks up and falls away, so that the sides are always covered by angular shingle forming slopes corresponding to the angle of friction of the rock in question. In the case of a volcanic mountain, however, the same form is due either to accumulation of fragmental material piled around the cup-shaped hollow, or crater, which is usually placed at the apex of the cone, and owing to which it is bluntly terminated, or else to the welling up from beneath of viscous matter in the manner presently to be described.

Views of Sir Humphrey Davy and L. von Buch.--The question how a volcanic cone came to be formed was not settled without a long controversy carried on by several naturalists of eminence. Some of the earlier writers of modern times on the subject of vulcanicity--such as Sir Humphrey Davy and Leopold von Buch--maintained that the conical form was due to upheaval by a force acting from below at a central focus, whereby the materials of which the mountain is formed were forced to assume a qu?qu?versal position--that is, a position in which the materials dip away from the central focus in every direction. But this view, originally contested by Scrope and Lyell, has now been generally abandoned. It will be seen on reflection that if a series of strata of ashes, tuff, and lava, originally horizontal, or nearly so, were to be forced upwards into a conical form by a central force, the result would be the formation of a series of radiating fissures ever widening from the circumference towards the focus. In the case of a large mountain such fissures, whether filled with lava or otherwise, would be of great breadth towards the focus, or central crater, and could not fail to make manifest beyond dispute their mechanical origin. But no fissures of the kind here referred to are, as a matter of fact, to be observed. Those which do exist are too insignificant and too irregular in direction to be ascribed to such an origin; so that the views of Von Buch and Davy must be dismissed, as being unsupported by observation, and as untenable on dynamical grounds. As a matter of fact, the "elevatory theory," or the "elevation-crater theory," as it is called by Scrope, has been almost universally abandoned by writers on vulcanicity.

Principal Varieties of Volcanic Mountains as regards Form.--But whilst rejecting the "elevatory theory," it is necessary to bear in mind that volcanic cones and dome-shaped elevations have been formed in several distinct ways, giving rise to varieties of structure essentially different. Two of the more general of these varieties of form, the crater-cone and the dome, are found in some districts, as in Auvergne, side by side. The crater-cone consists of beds or sheets of ashes, lapilli, and slag piled up in a conical form, with a central crater (or cup) containing the principal pipe through which these materials have been erupted; the dome, of a variety of trachytic lava, which has been extruded in a molten, or viscous, condition from a central pipe, and in such cases there is no distinct crater. There are other forms of volcanic mountains, such as those built up of basaltic matter, of which I shall have to speak hereafter, but the two former varieties are the most prevalent; and we may now proceed to consider the conditions under which the crater-cone volcanoes have been formed.

Crateriform Volcanic Cones.--Of this class nearly all the active volcanoes of the Mediterranean region--Etna, Vesuvius, Stromboli, and the Lipari Islands-- may be considered as representatives. They consist essentially of masses of fragmental material, which have from time to time been blown out of an orifice and piled up around with more or less regularity (according to the force exerted, and direction of the prevalent winds), alternating with sheets of lava. In this way mountains several thousand feet in height and of vast horizontal extent are formed. The fragmental materials thus accumulated are of all sizes, from the finest dust up to blocks many tons in weight, the latter being naturally piled around nearest to the orifice. The fine dust, blown high into the air by the explosive force of the gases and vapours, is often carried to great distances by the prevalent winds. Thus during the eruption of Vesuvius in A.D. 472 showers of ashes, carried high into the air by the westerly wind, fell over Constantinople at a distance of 750 miles.[3]

These loose, or partially consolidated, fragmental materials are rudely stratified, and slope downwards and outwards from the edge of the crater, so as to present the appearance of what is known as "the dip" of stratified deposits which have been upraised from the horizontal position by terrestrial forces. It was this excentrical arrangement which gave rise to the supposition that such volcanic ash-beds had been tilted up by a force acting in the direction of the volcanic throat, or orifice of eruption. The interior wall of Monte di Somma, the original crater of Vesuvius, presents a good illustration of such fragmental beds. I shall have occasion further on to describe more fully the structure of this remarkable mountain; so that it will suffice to say here that this old prehistoric crater, the walls of which enclose the modern cone of Vesuvius, is seen to be formed of irregular beds of ash, scori and fragmental masses, traversed by numerous dykes of lava, and sloping away outwards towards the surrounding plains.

Of similar materials are the flanks of Etna composed, even at great distances from the central crater; the beds of ash and agglomerate sometimes alternating with sheets of solidified lava and traversed by dykes of similar material of later date, injected from below through fissures formed during periods of eruptive energy. Numerous similar examples are to be observed in the Auvergne region of Central France and the Eifel. And here we find remarkable cases of "breached cones," or craters, which will require some special description. Standing on the summit of the Puy de De, and looking northwards or southwards, the eye wanders over a tract formed of dome-shaped hills and of extinct crater-cones rising from a granitic platform. But what is most peculiar in the scene is the ruptured condition of a large number of the cones with craters. In such cases the wall of the crater has been broken down on one side, and we observe that a stream of lava has been poured out through the breach and overflowed the plain below. The cause of this breached form is sufficiently obvious. In such cases there has been an explosion of ashes, stones, and scori?from the volcanic throat, by which a cone-shaped hill with a crater has been built up. This has been followed by molten lava welling up through the throat, and gradually filling the crater. But, as the lava is much more dense than the material of which the crater wall is composed, the pressure of the lava outwards has become too great for the resistance of the wall, which consequently has given way at its weakest part and, a breach being formed, the molten matter has flowed out in a stream which has inundated the country lying at the base of the cone. In one instance mentioned by Scrope, the original upper limit of the lake of molten lava has left its mark in the form of a ring of slag on the inside of the breached crater.[4]

Craterless Domes.--These differ essentially both in form and composition from those just described, and have their typical representatives in the Auvergne district, though not without their analogues elsewhere, as in the case of Chimborazo, in South America, one of the loftiest volcanic mountains in the world.

Taking the Puy de De, Petit Suchet, Cliersou, Grand Sarcoui in Auvergne, and the Mamelon in the Isle of Bourbon as illustrations, we have in all these cases a group of volcanic hills, dome-shaped and destitute of craters, the summits being rounded or slightly flattened. We also observe that the flanks rise more abruptly from their bases, and contrast in outline with the graceful curve of the crater cones. The dome-shaped volcanoes are generally composed of felsitic matter, whether domite, trachyte, or andesite, which has been extruded in a molten or viscous condition from some orifice or fissure in the earth's crust, and being piled up and spreading outwards, necessarily assumes such a form as that of a dome, as has been shown by experiment on a small scale by Dr. E. Reyer, of Graz.[5] The contrast between the two forms (those of the dome and the crater-cone) is exemplified in the case of the Grand Sarcoui and its neighbours. The former is composed of a species of trachyte; the latter of ashes and fragmental matter which have been blown out of their respective vents of eruption into the air, and piled up and around in a crateriform manner with sides of gradually diminishing slope outwards, thus giving rise to the characteristic volcanic curve. The two varieties here referred to, contrasting in form, composition, and colour of material, can be clearly recognised from the summit of the Puy de De, which rises by a head and shoulders above its fellows, and thus affords an advantageous standpoint from which to compare the various forms of this remarkable group of volcanic mountains.

Cotopaxi (Fig. 2) has been generally supposed to be a dome; but Whymper, who ascended the mountain in 1880, shows that it is a cone with a crater, 2,300 feet in largest diameter. He determined the height to be 19,613 feet above the ocean. Its real elevation above the sea is somewhat masked, owing to the fact that it rises from the high plain of Tapia, which is itself 8,900 feet above the sea surface. The smaller peak on the right (Fig. 2) is that of Carihuairazo, which reaches an elevation of over 16,000 feet.

Chimborazo, in Columbia, province of Quito, is one of the loftiest of the chain of the Andes, and is situated in lat. 1?30' S., long. 78?58' W. Though not in a state of activity, it is wholly composed of volcanic material, and reaches an elevation of over 20,000 feet above the ocean; its sides being covered by a sheet of permanent snow to a level of 2,600 feet below the summit.[6] Seen from the shores of the Pacific, after the long rains of winter, it presents a magnificent spectacle, "when the transparency of the air is increased, and its enormous circular summit is seen projected upon the deep azure blue of the equatorial sky. The great rarity of the air through which the tops of the Andes are seen adds much to the splendour of the snow, and aids the magical effect of its reflection."

Chimborazo was ascended by Humboldt and Bonpland in 1802 almost to the summit; but at a height of 19,300 feet by barometrical measurement, their further ascent was arrested by a wide chasm. Boussingault, in company with Colonel Hall, accomplished the ascent as far as the foot of the mass of columnar "trachyte," the upper surface of which, covered by a dome of snow, forms the summit of the mountain. The whole mass of the mountain consists of volcanic rock, varieties of andesite; there is no trace of a crater, nor of any fragmental materials, such as are usually ejected from a volcanic vent of eruption.[7]

Lava Crater-Cones.--A third form of volcanic mountain is that which has been built up by successive eruptions of basic lava, such as basalt or dolerite, when in a molten condition. These are very rare, and the slope of the sides depends on the amount of original viscosity. Where the lava is highly fused its slope will be slight, but if in a viscous condition, successive outpourings from the orifice, unable to reach the base of the mountain, will tend to form a cone with increasing slope upwards. Mauna Loa and Kilauea, in the Hawaiian Group, according to Professor J. D. Dana, are basalt volcanoes in a normal state. They have distinct craters, and the material of which the mountain is formed is basalt or dolerite. The volcano of Rangitoto in Auckland, New Zealand, appears to belong to this class.

Basalt is the most fusible of volcanic rocks, owing to the augite and magnetite it contains, so that it spreads out with a very slight slope when highly fused. Trachyte, on the other hand, is the least fusible owing to the presence of orthoclase felspar, or quartz; so that the volcanic domes formed of this material stand at a higher angle from the horizon than those of basaltic cones.

[1] Scenery and Geology of Scotland (1865), p. 214.

[2] Humboldt says: "The form of isolated conical mountains, as those of Vesuvius, Etna, the Peak of Teneriffe, Tunguagua, and Cotopaxi, is certainly the shape most commonly observed in volcanoes all over the globe."--Views of Nature, translated by E. C. Ott?and H. G. Bohn (1850).

[3] It is supposed that after the disastrous explosion of Krakatoa in 1883 the fine dust carried into the higher regions of the atmosphere was carried round almost the entire globe, and remained suspended for a lengthened period, as described in a future page.

[4] Another remarkable case is mentioned and figured by Judd, where one of the Lipari Isles, composed of pumice and rising out of the Mediterranean, has been breached by a lava-stream of obsidian.--Loc. cit., p. 123.

[5] Reyer has produced such dome-shaped masses by forcing a quantity of plaster of Paris in a pasty condition up through an orifice in a board; referred to by Judd, loc. cit., p. 125.

[6] Whymper determined the height to be 20,498 feet; Reiss and Stel make it 20,703 feet. Whymper thinks there may be a crater concealed beneath the dome of snow.--Travels amongst the Great Andes of the Equator, by Edward Whymper (1892).

[7] Whymper states that there is a prevalent idea that Cotopaxi and a volcano called Sangai act as safety-valves to each other. Sangai reaches an elevation (according to Reiss and Stel) of 17,464 feet, and sends intermittent jets of steam high into the air, spreading out into vast cumulus clouds, which float away southwards, and ultimately disappear.--Ibid., p. 73.

CHAPTER III.

LINES AND GROUPS OF ACTIVE VOLCANIC VENTS.

The globe is girdled by a chain of volcanic mountains in a state of greater or less activity, which may perhaps be considered a girdle of safety for the whole world, through which the masses of molten matter in a state of high pressure beneath the crust find a way of escape; and thus the structure of the globe is preserved from even greater convulsions than those which from time to time take place at various points on its surface. This girdle is partly terrestrial, partly submarine; and commencing at Mount Erebus, near the Antarctic Pole, ranging through South Shetland Isle, Cape Horn, the Andes of South America, the Isthmus of Panama, then through Central America and Mexico, and the Rocky Mountains to Kamtschatka, the Aleutian Islands, the Kuriles, the Japanese, the Philippines, New Guinea, and New Zealand, reaches the Antarctic Circle by the Balleny Islands. This girdle sends off branches at several points. (See Map, p. 23.)

(a.) The linear arrangement of active or dormant volcanic vents has been pointed out by Humboldt, Von Buch, Daubeny, and other writers. The great range of burning mountains of the Andes of Chili, Peru, Bolivia, and Mexico, that of the Aleutian Islands, of Kamtschatka and the Kurile Islands, extending southwards into the Philippines, and the branching range of the Sunda Islands are well-known examples. That of the West Indian Islands, ranging from Grenada through St. Vincent, St. Lucia, Martinique, Dominica, Guadeloupe, Montserrat, Nevis, and St. Eustace,[1] is also a remarkable example of the linear arrangement of volcanic mountains. On tracing these ranges on a map of the world[2] (Map, p. 23), it will be observed that they are either strings of islands, or lie in proximity to the ocean; and hence the view was naturally entertained by some writers that oceanic water, or at any rate that of a large lake or sea, was a necessary agent in the production of volcanic eruptions. This view seems to receive further corroboration from the fact that the interior portions of the continents and large islands such as Australia are destitute of volcanoes in action, with the remarkable exceptions of Mounts Kenia and Kilimanjaro in Central Africa, and a few others. It is also very significant in this connection that many of the volcanoes now extinct, or at least dormant, both in Europe and Asia, appear to have been in proximity to sheets of water during the period of activity. Thus the old volcanoes of the Haur 鈔, east of the Jordan, appear to have been active at the period when the present Jordan valley was filled with water to such an extent as to constitute a lake two hundred miles in length, but which has now shrunk back to within the present limits of the Dead Sea.[3] Again, at the period when the extinct volcanoes of Central France were in active operation, an extensive lake overspread the tract lying to the east of the granitic plateau on which the craters and domes are planted, now constituting the rich and fertile plain of Clermont.

Such instances are too significant to allow us to doubt that water in some form is very generally connected with volcanic operations; but it does not follow that it was necessary to the original formation of volcanic vents, whether linear or sporadic. If this were so, the extinct volcanoes of the British Isles would still be active, as they are close to the sea-margin, and no volcano would now be active which is not near to some large sheet of water. But Jorullo, one of the great active volcanoes of Mexico, lies no less than 120 miles from the ocean, and Cotopaxi, in Ecuador, is nearly equally distant. Kilimanjaro, 18,881 feet high, and Kenia, in the equatorial regions of Central

Africa, are about 150 miles from the Victoria Nyanza, and a still greater distance from the ocean; and Mount Demavend, in Persia, which rises to an elevation of 18,464 feet near the southern shore of the Caspian Sea, a volcanic mountain of the first magnitude, is now extinct or dormant.[4] Such facts as these all tend to show that although water may be an accessory of volcanic eruptions, it is not in all cases essential; and we are obliged, therefore, to have recourse to some other theory of volcanic action differing from that which would attribute it to the access of water to highly heated or molten matter within the crust of the earth.

(b.) Leopold von Buch on Rents and Fissures in the Earth's Crust.--The view of Leopold von Buch, who considered that the great lines of volcanic mountains above referred to rise along the borders of rents, or fissures, in the earth's crust, is one which is inherently probable, and is in keeping with observation. That the crust of the globe is to a remarkable extent fissured and torn in all directions is a phenomenon familiar to all field geologists. Such rents and fissures are often accompanied by displacement of the strata, owing to which the crust has been vertically elevated on one side or lowered on the other, and such displacements (or "faults") sometimes amount to thousands of feet. It is only occasionally, however, that such fractures are accompanied by the extrusion of molten matter; and in the North of England and Scotland dykes of igneous rock, such as basalt, which run across the country for many miles in nearly straight lines, often cut across the faults, and are only rarely coincident with them. Nevertheless, it can scarcely be a question that the grand chain of volcanic mountains which stretches almost continuously along the Andes of South America, and northwards through Mexico, has been piled up along the line of a system of fissures in the fundamental rocks parallel to the coast, though not actually coincident therewith.

(c.) The Cordilleras of Quito.--The structure and arrangement of the Cordilleras of Quito, for example, are eminently suggestive of arrangement along lines of fissure. As shown by Alexander von Humboldt,[5] the volcanic mountains are disposed in two parallel chains, which run side by side for a distance of over 500 miles northwards into the State of Columbia, and enclose between them the high plains of Quito and Lacunga. Along the eastern chain are the great cones of El Altar, rising to an elevation of 16,383 feet above the ocean, and having an enormous crater apparently dormant or extinct, and covered with snow; then Cotopaxi (Fig. 2), its sides covered with snow, and sending forth from its crater several columns of smoke; then Guamani and Cayambe (19,000 feet), huge truncated cones apparently extinct; these constitute the eastern chain of volcanic heights. The western chain contains even loftier mountains. Here we find the gigantic Chimborazo, an extinct volcano whose summit is white with snow; Carihuairazo[6] and Illiniza, a lofty pointed peak like the Matterhorn; Corazon, a snow-clad dome, reaching a height of 15,871 feet; Atacazo and Pichincha, the latter an extinct volcano reaching an elevation of 15,920 feet; such is the western chain, remarkable for its straightness, the volcanic cones being planted in one grand procession from south to north. This rectilinear arrangement of the western chain, only a little less conspicuous in the eastern, is very suggestive of a line of fracture in the crust beneath. And when we contemplate the prodigious quantity of matter included within the limits of these colossal domes and their environments, all of which has been extruded from the internal reservoirs, we gain some idea of the manner in which the contracting crust disposes of the matter it can no longer contain.[7]

Between the volcanoes of Quito and those of Peru there is an intervening space of fourteen degrees of latitude. This is occupied by the Andes, regarding the structure of which we have not much information except that at this part of its course it is not volcanic. But from Arequipa in Peru (lat. 16?S.), an active volcano, we find a new series of volcanic mountains continued southwards through Tacora (19,740 feet), then further south the more or less active vents of Sajama (22,915 feet), Coquina, Tutupaca, Calama, Atacama, Toconado, and others, forming an almost continuous range with that part of the desert of Atacama pertaining to Chili. Through this country we find the volcanic range appearing at intervals; and still more to the southwards it is doubtless connected with the volcanoes of Patagonia, north of the Magellan Straits, and of Tierra del Fuego. Mr. David Forbes considers that this great range of volcanic mountains, lying nearly north and south, corresponds to a line of fracture lying somewhat to the east of the range.[8]

(d.) Other Volcanic Chains.--A similar statement in all probability applies to the systems of volcanic mountains of the Aleutian Isles, Kamtschatka, the Kuriles, the Philippines, and Sunda Isles. Nor can it be reasonably doubted that the western American coast-line has to a great extent been determined, or marked out, by such lines of displacement; for, as Darwin has shown, the whole western coast of South America, for a distance of between 2000 and 3000 miles south of the Equator, has undergone an upward movement in very recent times--that is, within the period of living marine shells--during which period the volcanoes have been in activity.[9]

(e.) The Kurile Islands.--This chain may also be cited in evidence of volcanic action along fissure lines. It connects the volcanoes of Kamtschatka with those of Japan, and the linear arrangement is apparent. In the former peninsula Erman counted no fewer than thirteen active volcanic mountains rising to heights of 12,000 to 15,000 feet above the sea.[10] In the chain of the Kuriles Professor John Milne counted fifty-two well-defined volcanoes, of which nine, perhaps more, are certainly active.[11] They are not so high as those of Kamtschatka; but, on the other hand, they rise from very deep oceanic waters, and have been probably built up from the sea bottom by successive eruptions of tuff, lava, and ash. According to the view of Professor Milne, the volcanoes of the Kurile chain are fast becoming extinct.

(f.) Volcanic Groups.--Besides the volcanic vents arranged in lines, of which we have treated above, there are a large number, both active and extinct, which appear to be disposed in groups, or sporadically distributed, over various portions of the earth's surface. I say appear to be, because this sporadic distribution may really be resolvable (at least in some cases) into linear distribution for short distances. Thus the Neapolitan Group, which might at first sight seem to be arranged round Vesuvius as a centre, really resolves itself into a line of active and extinct vents of eruption, ranging across Italy from the Tyrrhenian Sea to the Adriatic, through Ischia, Procida, Monte Nuovo and the Phlegren Fields, Vesuvius, and Mount Vultur.[12] Again, the extinct volcanoes of Central France, which appear to form an isolated group, indicate, when viewed in detail, a linear arrangement ranging from north to south.[13] Another region over which extinct craters are distributed lies along the banks of the Rhine, above Bonn and the Moselle; a fourth in Hungary; a fifth in Asia Minor and Northern Palestine; and a sixth in Central Asia around Lake Balkash. These are all continental, and the linear distribution is not apparent.

[1] For an interesting account of this range of volcanic islands see Kingsley's At Last. The grandest volcanic peak is that of Guadeloupe, rising to a height of 5000 feet above the ocean, amidst a group of fourteen extinct craters. But

the most active vent of the range is the Souffrie of St. Vincent. In the eruption of 1812 this mountain sent forth clouds of pumice, scori and ashes, some of which were carried by an upper counter current to Barbados, one hundred miles to the eastward, covering the surface with volcanic dust to a depth of several inches.

[2] An excellent, and perhaps the most recent, map of this kind is that given by Professor Prestwich in his Geology, vol. i. p. 216. One on a larger scale is that by Keith Johnston in his Physical Atlas.

[3] Memoir on the Physical Geology and Geography of Arabia Petra, Palestine, etc., published for the Committee of the Palestine Exploration Fund (1886), p. 113, etc.

[4] This mountain was ascended in 1837 by Mr. Taylor Thomson, who found the summit covered with sulphur, and from a cone fumes at a high temperature issued forth, but there was no eruption.--Journ. Roy. Geographical Soc., vol. viii. p. 109.

[5] Humboldt, Atlas der Kleineren Schriften (1853).

[6] Ascended by Whymper June 29, 1880. He found the elevation to be 16,515 feet.

[7] The arrangement of the volcanoes of Peru and Bolivia is also suggestive of a double line of fissure, while those of Chili suggest one single line. The volcanoes of Arequipa, in the southern part of Peru, are dealt with by Dr. F. H. Hatch, in his inaugural dissertation, Ueber die Gesteine der Vulcan-Gruppe von Arequipa (Wien, 1886). The volcanoes rise to great elevations, having their summits capped by snow. The volcano of Charchani, lying to the north of Arequipa, reaches an elevation of 18,382 Parisian feet. That of Pichupichu reaches a height of 17,355 Par. feet. The central cone of Misti has been variously estimated to range from 17,240 to 19,000 Par. feet. The rocks of which the mountains are composed consist of varieties of andesite.

[8] D. Forbes, "On the Geology of Bolivia and Southern Peru," Quarterly Journal of the Geological Society, vol. xvii. p. 22 (1861).

[9] Darwin, Structure and Distribution of Coral Reefs, second edition, p. 186.

[10] Erman, Reise um die Welt.

[11] Milne, "Cruise amongst the Kurile Islands," Geol. Mag., New Ser. (August 1879).

[12] See Daubeny, Volcanoes, Map I.

[13] Sir A. Geikie has connected as a line of volcanic vents those of Sicily, Italy, Central France, the N. E. of Ireland, the Inner Hebrides and Iceland, of which the central vents are extinct or dormant, the extremities active.

CHAPTER IV.

MID-OCEAN VOLCANIC ISLANDS.

Oceanic Islands.--By far the most extensive regions with sporadically distributed volcanic vents, both active and extinct, are those which are overspread by the waters of the ocean, where the vents emerge in the form of islands. These are to be found in all the great oceans, the Atlantic, the Pacific, and the Indian; but are especially numerous over the central Pacific region. As Kotzebue and subsequently Darwin have pointed out, all the islands of the Pacific are either coral-reefs or of volcanic origin; and many of these rise from great depths; that is to say, from depths of 1000 to 2000 fathoms. It is unnecessary here to attempt to enumerate all these islands which rise in solitary grandeur from the surface of the ocean, and are the scenes of volcanic operations; a few may, however, be enumerated.

(a.) Iceland.--In the Atlantic, Iceland first claims notice, owing to the magnitude and number of its active vents and the variety of the accompanying phenomena, especially the geysers. As Lyell has observed,[1] with the exception of Etna and Vesuvius, the most complete chronological records of a series of eruptions in existence are those of Iceland, which come down from the ninth century of our era, and which go to show that since the twelfth century there has never been an interval of more than forty years without either an eruption or a great earthquake. So intense is the volcanic energy in this island that some of the eruptions of Hecla have lasted six years without cessation. Earthquakes have often shaken the whole island at once, causing great changes in the interior, such as the sinking down of hills, the rending of mountains, the desertion by rivers of their channels, and the appearance of new lakes. New islands have often been thrown up near the coast, while others have disappeared. In the intervals between the eruptions, innumerable hot springs afford vent to the subterranean heat, and solfataras discharge copious streams of inflammable matter. The volcanoes in different parts of the island are observed, like those of the Phlegren Fields, to be in activity by turns, one vent serving for a time as a safety-valve for the others. The most memorable eruption of recent years was that of Skapt Jokul in 1783, when a new island was thrown up, and two torrents of lava issued forth, one 45 and the other 50 miles in length, and which, according to the estimate of Professor Bischoff, contained matter surpassing in magnitude the bulk of Mont Blanc. One of these streams filled up a large lake, and, entering the channel of the Skapt? completely dried up the river. The volcanoes of Iceland may be considered as safety-valves to the region in which lie the British Isles.

(b.) The Azores, Canary, and Cape de Verde Groups.--This group of volcanic isles rises from deep Atlantic waters north of the Equator, and the vents of eruption are partially active, partially dormant, or extinct. It must be supposed, however, that at a former period volcanic action was vastly more energetic than at present; for, except at the Grand Canary, Gomera, Forta Ventura, and Lancerote, where various non-volcanic rocks are found, these islands appear to have been built up from their foundations of eruptive materials. The highest point in the Azores is the Peak of Pico, which rises to a height of 7016 feet above the ocean. But this great elevation is surpassed by that of the Peak of Teneriffe (or Pic de Teyde) in the Canaries, which attains to an elevation of 12,225 feet, as determined by Professor Piazzi Smyth.[2]

This great volcanic cone, rising from the ocean, its summit shrouded in snow, and often protruding above the clouds, must be an object of uncommon beauty and interest when seen from the deck of a ship. (Fig. 4.) The central cone, formed of trachyte, pumice, obsidian, and ashes, rises out of a vast caldron of older basaltic rocks with precipitous inner walls--much as the cone of Vesuvius rises from within the partially encircling walls of Somma. (Fig. 5.) From the summit issue forth sulphurous vapours, but no flame.

Piazzi Smyth, who during a prolonged visit to this mountain in 1856 made a careful survey of its form and structure, shows that the great cone is surrounded by an outer ring of basalt enclosing two foci of eruption, the lavas from which have broken through the ring of the outer crater on the western side, and have poured down the mountain. At the top of the peak its once active crater is filled up, and we find a convex surface ("The Plain of Rambleta") surmounted towards its eastern end by a diminutive cone, 500 feet high, called "Humboldt's Ash Cone." The slope of the great cone of Teneriffe ranges from 28 to 38 and below a level of 7000 feet the general slope of the whole mountain down to the water's edge varies from 10?to 12?from the horizontal. The great cone is penetrated by numerous basaltic dykes.

The Cape de Verde Islands, which contain beds of limestone along with volcanic matter, possess in the island of Fuego an active volcano, rising to a height of 7000 feet above the surface of the ocean. The central cone, like that of Teneriffe, rises from within an outer crater, formed of basalt alternating with beds of agglomerate, and traversed by numerous dykes of lava. This has been broken down on one side like that of Somma; and over its flanks are scattered numerous cones of scori? the most recent dating from the years 1785 and 1799.[3]

The volcanoes of Lancerote have a remarkably linear arrangement from west to east across the island. They are not yet extinct; for an eruption in 1730 destroyed a large number of villages, and covered with lava the most fertile tracts in the island, which at the time of Leopold von Buch's visit lay waste and destitute of herbage.[4] In the island of Palma there is one large central crater, the Caldera de Palma, three leagues in diameter, the walls of which conform closely to the margin of the coast. Von Buch calls this crater "une merveille de la nature," for it distinguishes this isle from all the others, and renders it one of the most interesting and remarkable amongst the volcanic islands of the ocean. The outer walls are formed of basaltic sheets, and towards the south this great natural theatre is connected with the ocean by a long straight valley, called the "Barranco de los Dolores," along whose sides the structure of the mountain is deeply laid open to view. The outer flanks of the crater are furrowed by a great number of smaller barrancos radiating outward from the rim of the caldera. Von Buch regards the barrancos as having been formed during the upheaval of the island, according to his theory of the formation of such mountains (the elevation-theory); but unfortunately for his views, these ravines widen outwards from the centre, or at least do not become narrower in that direction, as would be the case were the elevation-theory sound. The maps which accompany Von Buch's work are remarkably good, and were partly constructed by himself.

(c.) Volcanic Islands in the Atlantic south of the Equator.--The island of Ascension, formed entirely of volcanic matter, rises from a depth of 2000 fathoms in the very centre of the Atlantic. As described by Darwin, the central and more elevated portions are formed of trachytic matter, with obsidian and laminated ash beds. Amongst these are found ejected masses of unchanged granite, fragments of which have been torn from the central pipe during periods of activity, and would seem to indicate a granitic floor, or at least an original floor upon which more recent deposits may have been superimposed. In St. Helena we seem, according to Daubeny, to have the mere wreck of one great crater, no one stream of lava being traceable to its source, while dykes of lava are scattered in profusion throughout the whole substance of the basaltic masses which compose the island. Tristan da Cunha, in the centre of the South Atlantic, rises abruptly from a depth of 12,150 feet, at a distance of 1500 miles from any land; and one of its summits reaches an elevation of 7000 feet, being a truncated cone composed of alternating strata of tuff and augitic lava, surrounding a crater in which is a lake of pure water. The volcano is extinct or dormant.

Were the waters of the ocean to be drawn off, these volcanic islands would appear like stupendous conical mountains, far loftier, and with sides more precipitous, than any to be found on our continental lands, all of which rise from platforms of considerable elevation. The enormous pressure of the water on their sides enables these mid-oceanic islands to stand with slopes varying from the perpendicular to a smaller extent than if they were sub-aerial; and it is on this account that we find them rising with such extraordinary abruptness from the "vasty deep."

(d.) Volcanic Islands of the Pacific.--The volcanic islands of this great ocean are scattered over a wide tract on both sides of the equator. Those to the north of this line include the Sandwich Islands, the Mariana or Ladrone Islands, South Island, and Bonin Sima; south of the equator, the Galapagos, New Britain, Salomon, Santa Cruz, New Hebrides, the Friendly and Society Isles. While the coral reefs and islands of the Pacific may be recognised by their slight elevation above the surface of the waters, those of volcanic origin and containing active or extinct craters of eruption generally rise into lofty elevations, so that the two kinds are called the Low Islands and High Islands respectively. Amongst the group are trachytic domes such as the Mountain of Tobreonu in the Society Islands, rising to a height probably not inferior to that of Etna, with extremely steep sides, and holding a lake on its summit.[5] The linear arrangement of some of the volcanic islands of the Pacific is illustrated by those of the Tonga, or Friendly, Group, lying to the north of New Zealand. They consist of three divisions--(1) the volcanic; (2) those formed of stratified volcanic tuff, sometimes entirely or partially covered by coralline limestone; and (3) those which are purely coralline. The first form a chain of lofty cones and craters, lying in a E.N.E. and W.S.W. direction, and rising from depths of over 1000 fathoms. Mr. J. J. Lister, who has described the physical characters of these islands, has shown very clearly that they lie along a line--probably that of a great fissure--stretching from the volcanic island of Amargura on the north (lat. 18?S.), through Lette, Metis, Kao (3030 feet), Tofua, Falcon, Honga Tonga, and the Kermadec Group into the New Zealand chain on the south. Some of these volcanoes are in a state of intermittent activity, as in the case of Tofua (lat. 20?30' S.), Metis Island, and Amargura; the others are dormant or extinct. The whole group appears to have been elevated at a recent period, as some of the beds of coral have been raised 1272 feet and upward above the sea-level, as in the case of Eua Island.[6] The greater number of the Pacific volcanoes appear to be basaltic; such as those of the Hawaiian Group, which have been so fully described by Professor J. D. Dana.[7] Here fifteen volcanoes of the first class have been in brilliant action; all of which, except three, are now extinct, and these are in Hawaii the largest and most eastern of the group. This island contains five volcanic mountains, of which Kea, 13,805 feet, is the highest; next to that, Loa, 13,675 feet; after these, Hualalai, rising 8273 feet; Kilauea, 4158 feet; and Kohala, 5505 feet above the sea; this last is largely buried beneath the lavas of Mauna Kea. The group contains a double line of volcanoes, one lying to the north and west of the other; and as the highest of the Hawaiian Group rises from a depth in the ocean of over 2000 fathoms, the total elevation of this mountain from its base on the bed of the ocean is not far from 26,000 feet, an elevation about that of the Himalayas. Mauna Kea has long been extinct, Hualalai has been dormant since 1801; but Mauna Loa is terribly active, there having been several eruptions, accompanied by earthquakes, within recent years, the most memorable being those of 1852 and 1868. In the former case the lava rose from the deep crater into "a lofty mountain," as described by Mr. Coan,[8] and then flowed away eastward for a distance of twenty miles. The interior of the crater consists of a vast caldron, surrounded by a precipice 200 to 400 feet in depth, with a circumference of about fifteen miles, and containing within it a second crater, bounded by a black ledge with a steep wall of basalt--a crater within a crater; and from the floor of the inner crater, formed of molten basalt, in a seething and boiling state, arise a large number of small cones and pyramids of lava, some emitting columns of grey smoke, others brilliant flames and streams of molten lava, presenting a wonderful spectacle, the effect of which is heightened by the constant roaring of the vast furnaces below.[9]

[1] Principles of Geology, 11th edition, vol. ii. p. 48.

[2] Smyth, Report on the Teneriffe Astronomical Experiment of 1856. Humboldt makes the elevation 12,090 feet. A beautiful model of the Peak was constructed by Mr. J. Nasmyth from Piazzi Smyth's plans, of which photographs are given by the latter.

[3] Daubeny, loc. cit., p. 460.

[4] Iles Canaries, p. 37.

[5] Daubeny, loc. cit., p. 426.

[6] Lister, "Notes on the Geology of the Tonga Islands," Quart. Jour. Geol. Soc., No. 188, p. 590 (1891).

[7] Dana, Characteristics of Volcanoes, with Contributions of Facts and Principles from the Hawaiian Islands. London, 1890.--Also, Geology of the American Exploring Expedition--Volcanoes of the Sandwich Islands.

[8] Coan, Amer. Jour. of Science, 1853.

[9] W. Ellis, the missionary, has given a vivid description of this volcano in his Tour of Hawaii. London, 1826.--Plans of the crater will be found in Professor Dana's work above quoted.

PART II.

EUROPEAN VOLCANOES.

CHAPTER I.

VESUVIUS.

Having now dealt in a necessarily cursory manner with volcanoes of distant parts of the globe, we may proceed to the consideration of the group of active volcanoes which still survive in Europe, as they possess a special interest, not only from their proximity and facility of access, at least to residents in Europe and the British Isles, but from their historic incidents; and in this respect Vesuvius, though not by any means the largest of the group, stands the first, and demands more special notice. The whole group rises from the shores of the Mediterranean, and consists of Vesuvius, Etna, Stromboli, one of the Lipari Islands, and Vulcano, a mountain which has given the name to all mountains of similar origin with itself.[1] Along with these are innumerable cones and craters of extinct or dormant volcanoes, of which a large number have been thrown out on the flanks of Etna.

(a.) Prehistoric Ideas regarding the Nature of this Mountain.--Down to the commencement of the Christian era this mountain had given no ostensible indication that it contained within itself a powerful focus of volcanic energy. True, that some vague tradition that the mountain once gave forth fire hovered around its borders; and several ancient writers, amongst them Diodorus Siculus and Strabo, inferred from the appearances of the higher parts of the mountain and the character of the rocks, which were "cindery and as if eaten by fire," that the country was once in a burning state, "being full of fiery abysses, though now extinct from want of fuel." Seneca (B.C. 1 to A.D. 64) had detected the true character of Vesuvius, as "having been a channel for the internal fire, but not its food;" nevertheless, at this period the flanks of the mountain were covered by fields and vineyards, while the summit, partially enclosed with precipitous walls of the long extinct volcano, Somma, was formed of slaggy and scoriaceous material, with probably a covering of scrub. Here it was that the gladiator Spartacus (B.C. 72), stung by the intolerable evils of the Roman Government, retreated to the very summit of the mountain with some trusty followers. Clodius the Pr 鎡 or, according to the narration of Plutarch, with a party of three thousand men, was sent against them, and besieged them in a mountain (meaning Vesuvius or Somma) having but one narrow and difficult passage, which Clodius kept guarded; all the rest was encompassed with broken and slippery precipices, but upon the top grew a great many wild vines; the besieged cut down as many as they had need of, and twisted them into ladders long enough to reach from thence to the bottom, by which, without any danger, all got down except one, who stayed behind to throw them their arms, after which he saved himself with the rest.[2] "On the top" must (as Professor Phillips observes) be interpreted the summit of the exterior slope or crater edge, which would appear from the narrative to have broken down on one side, affording an entrance and mode of egress by which Spartacus fell upon, and surprised, the negligent Clodius Glabrus.

In fancied security, villas, temples, and cities had been erected on the slopes of the mountain. Herculaneum, Pompeii, and Stabi? the abodes of art, luxury, and vice, had sprung up in happy ignorance that they "stood on a volcano," and that their prosperity was to have a sudden and disastrous close.[3]

(b.) Premonitory Earthquake Shocks.--The first monitions of the impending catastrophe occurred in the 63rd year after Christ, when the whole Campagna was shaken by an earthquake, which did much damage to the towns and villas surrounding the mountain even beyond Naples. This was followed by other shocks; and in Pompeii the temple of Isis was so much damaged as to require reconstruction, which was undertaken and carried out by a citizen at his own expense.[4] These earthquake shakings continued for sixteen years. At length, on the night of August 24th, A.D. 79, they became so violent that the whole region seemed to reel and totter, and all things appeared to be threatened with destruction. The next day, about one in the afternoon, there was seen to rise in the direction of Vesuvius a dense cloud, which, after ascending from the summit of the mountain into the air for a certain height in one narrow, vertical trunk, spread itself out laterally in such a form that the upper part might be compared to the cluster of branches, and the lower to the stem of the pine which forms so common a feature in the Italian landscape.[5]

(c.) Pliny's Letters to Tacitus.--For an account of what followed we are indebted to the admirable letters of the younger Pliny, addressed to the historian Tacitus, recounting the events which caused, or accompanied, the death of his uncle, the elder Pliny, who at the time of this first eruption of Vesuvius was in command of the Roman fleet at the entrance to the Bay of Naples. These letters, which are models of style and of accurate description, are too long to be inserted here; but he recounts how the dense cloud which hung over the mountain spread over the whole surrounding region, sometimes illuminated by flashes of light more vivid than lightning; how showers of cinders, stones, and ashes fell in such quantity that his uncle had to flee from Stabi? and that even at so great a distance as Misenum they encumbered the surface of the ground; how the ground heaved and the bed of the sea was upraised; how the cloud descended on Misenum, and even the island of Capre?was concealed from view; and finally, how, urged by a friend who had arrived from Spain, he, with filial affection, supported the steps of his mother in flying from the city of destruction. Such being the condition of the atmosphere and the effects of the eruption at a point so distant as Cape Misenum, some sixteen geographical miles from the focus of eruption, it is only to be expected that places not half the distance, such as Herculaneum, Pompeii, and even Stabi? with many villages and dwellings, should have shared a worse fate. The first of these cities, situated on the coast of the Bay of Naples, appears to have been overwhelmed by volcanic mud; Pompeii was buried in ashes and lapilli, and Stabi?probably shared a similar fate.[6]

(d.) Appearance of the Mountain at the Commencement of the Christian Era.--At the time of the first recorded eruption Vesuvius appears to have consisted of only a single cone with a crater, now known as Monte di Somma, the central cone of eruption which now rises from within this outer ruptured casing not having been formed. (Fig. 6.) The first effect of the eruption of the year 79 was to blow out the solidified covering of slag and scori?forming the floor of the caldron. Doubtless at the close of the eruption a cone of fragmental matter and lava of some slight elevation was built up, and, if so, was subsequently destroyed; for, as we shall presently see by the testimony of the Abate Guilio Cesare Braccini, who examined the mountain not long before the great eruption of A.D. 1631, there was no central cone to the mountain at that time; and the mountain had assumed pretty much the appearance it had at the time that Spartacus took refuge within the walls of the great crater.

(e.) Destruction of Pompeii.--Pompeii was overwhelmed with dry ashes and lapilli. Sir W. Hamilton found some of the stones to weigh eight pounds. At the time of the author's visit, early in April 1872, the excavations had laid open a section about ten feet deep, chiefly composed of alternating layers of small pumice stones (lapilli) and volcanic dust. It was during the sinking of a well in 1713 upon the theatre containing the statues of Hercules and Cleopatra that the existence of the ancient city was accidentally discovered.

(f.) More recent eruptions.--Since the first recorded eruption in A.D. 79 down to the present day, Vesuvius has been the scene of numerous intermittent eruptions, of which some have been recorded; but many, doubtless, are forgotten.

In A.D. 203, during the reign of Severus, an eruption of extraordinary violence took place, which is related by Dion Cassius, from whose narrative we may gather that at this time there was only one large crater, and that the central cone of Vesuvius had not as yet been upraised. In A.D. 472 an eruption occurred of such magnitude as to cover all Europe with fine dust, and spread alarm even at Constantinople.

(g.) Eruption of 1631.--In December 1631 occurred the great convulsion whose memorials are written widely on the western face of Vesuvius in ruined villages. This eruption left layers of ashes over hundreds of miles of country, or heaps of mud swept down by hot water floods from the crater; the crater itself having been dissipated in the convulsion. Braccini, who examined the mountain not long before this eruption, found apparently no cone (or mount) like that of the modern Vesuvius. He states that the crater was five miles in circumference, about a thousand paces deep (or in sloping descent), and its sides covered with forest trees and brushwood, while at the bottom there was a plain on which cattle grazed.[7] It would seem that the mountain had at this time enjoyed a long interval of rest, and that it had reverted to very much the same state in which it was at the period of the first eruption, when the flanks were peopled by inhabitants living in fancied security. But six months of violent earthquakes, which grew more violent towards the close of 1631, heralded the eruption which took place in December, accompanied by terrific noises from within the interior of the mountain. The inhabitants of the coast were thus warned of the approaching danger, and had several days to arrange for their safety; but in the end, a great part of Torre del Greco was destroyed, and a like fate overtook Resina and Granatello, with a loss of life reported at 18,000 persons. During the eruption clouds condensed into tempests of rain, and hot water from the mountain, forming deluges of mud, swept down the sides, and reached even to Nola and the Apennines. Nor was the sea unmoved. It retired during the violent earthquakes, and then returned full thirty paces beyond its former limits.

Not indeed until near the close of the seventeenth century is there any evidence that the central cone of Vesuvius was in existence; but in October 1685 an eruption occurred which is recorded by Sorrentino, during which was erected "a new mountain within, and higher than the old one, and visible from Naples," a statement evidently referable to the existing cone--so that it is little more than two centuries since this famous volcanic mountain assumed its present form.

(h.) Eruptions between the years 1500 and 1800.--Since A.D. 1500 there have been fifty-six recorded eruptions of Vesuvius; one of these in 1767 was of terrific violence and destructiveness, and is represented by Sir William Hamilton in views taken both before and during the eruption. A pen-and-ink drawing of the appearance of the crater before the eruption is here reproduced from Hamilton's picture, from which it will be seen that the central crater contained within itself a second crater-cone, from whence steam, lava, and stones were being erupted (Fig. 7). Thus it will be seen that Vesuvius at this epoch consisted of three crater-cones within each other. The first, Monte di Somma; the second, the cone of Vesuvius; and the third, the little crater-cone within the second. During this eruption, vast lava-sheets invaded the fields and vineyards on the flanks of the mountain. A vivid account of this eruption, as witnessed by Padre Torre, is given by Professor Phillips.[8] We shall pass over others without further reference until we come down to our own times, in which Vesuvius has resumed its old character, and in one grand exhibition of volcanic energy, which took place in 1872, has evinced to the world that it still contains within its deep-seated laboratory all the elements of destructive force which it exhibited at the commencement of our era.

(i.) Structure of the Neapolitan Campagna.--But before giving a description of this terrific outburst of volcanic energy, it may be desirable to give some account of the physical position and structure of this mountain, by which the phenomena of the eruption will be better understood. Vesuvius and the Neapolitan Campagna are formed of volcanic materials bounded on the west by the Gulf of Naples, and on the east and south by ranges of Jurassic limestone, a prolongation of the Apennines, which send out a spur bounding the bay on the south, and forming the promontory of Sorrento. The little island of Capri is also formed of limestone, and is dissevered from the promontory by a narrow channel. The northern side of the bay is, however, formed of volcanic materials; it includes the Phlegren Fields (Campi Phlegr), and terminates in the promontory of Miseno. Lying in the same direction are the islands of Procida and Ischia, also volcanic. Hence it will be seen that the two horns of the bay are formed of entirely different materials, that of Miseno on the north being volcanic, that of Sorrento on the south being composed of Jurassic limestone, of an age vastly more ancient than the volcanic rocks on the opposite shore. (Map, p. 52.)

The general composition of the Neapolitan Campagna, from which the mountain rises, has been revealed by means of the Artesian well sunk to a depth of about 500 metres (1640 feet) at the Royal Palace of Naples, and may be generalised as follows:--

{ Recent beds of volcanic tuff (1) From surface to depth of { with marine shells, and containing 715 feet { fragments of trachytic { lava, etc. (Volcanic Beds).

{ Bituminous sands and marls (2) From 715 to 1420 { with marine shells of recent { species(?) (Pre-Volcanic Beds).

(3) From 1420 to 1574 { EOCENE BEDS. Micaceous sandstone { and marl (Macigno).

(4) From 1574 to bottom { JURASSIC BEDS. Apennine { Limestone.

From the above section, for which we are indebted to Mr. Johnston-Lavis, the most recent writer on Vesuvius, it would appear that the first volcanic explosions by which the mountain was ultimately to be built up took place after the deposition of the sands and marls (No. 2), while the whole Campagna was submerged under the waters of the Mediterranean. By the accumulation of the stratified tuff (No. 1), the sea-bed was gradually filled up during a period of volcanic activity, and afterwards elevated into dry land.[9]

(j.) Present Form and Structure of Vesuvius and Somma.--The outer cone of Vesuvius, or Monte di Somma, rises from a circular platform by a moderately gentle ascent, and along the north and east terminates in a craggy crest, with a precipitous cliff descending into the Atria del Cavallo, forming the wall of the ancient crater throughout half its circumference; this wall is formed of scori? ashes, and lapilli, and is traversed by numerous dykes of lava. Along the west and south this old crater has been broken down; but near the centre there remains a round-backed ridge of similar materials, once doubtless a part of the original crater of Somma, rising above the slopes of lava on either hand. On this has been erected the Royal Observatory, under the superintendence of Professor Luigi Palmieri, where continuous observations are being made, by means of delicate seismometers, of the earth-tremors which precede or accompany eruptions; for it is only justice to say that Vesuvius gives fair warning of impending mischief, and the instruments are quick to notify any premonitory symptoms of a coming catastrophe. The elevation of the Observatory is 2080 feet above the sea.

On either side of the Observatory ridge are wide channels filled to a certain height with lavas of the nineteenth and preceding centuries, the most recent presenting an aspect which can only be compared to a confused multitude of black serpents and pachyderms writhing and interlocked in some frightful death-struggle. Some of this lava, ten years old, as we cross its rugged and black surface presents gaping fissures, showing the mass to be red-hot a few feet from the surface, so slow is the process of cooling. These lava-streams-- some of them reaching to the sea-coast--have issued forth from the Atria at successive periods of eruption.

From the midst of the Atria rises the central cone, formed of cinders, scori? and lava-streams, and fissured along lines radiating from the axis. This cone is very steep, the angle being about 40?45?from the horizontal, and is formed of loose cindery matter which gives way at every step, and is rather difficult to climb. But on reaching the summit we look down into the crater, displaying a scene of ever-varying characters, rather oval in form, and about 1100 yards in diameter. From the map of Professor Guiscardi, published in 1855, there are seen two minor craters within the central one, formed in 1850, and an outflow of lava from the N.W. down the cone. At the time of the author's visit the crater was giving indications, by the great quantity of sulphurous gas and vapour rising from its surface, and small jets of molten lava beginning to flow down the outer side, of the grand outburst of internal forces which was presently to follow.

(k.) Eruption of 1872.--The grand eruption of 1872, of which a detailed account is given by Professor Palmieri,[10] commenced with a slight discharge of incandescent projectiles from the crater; and on the 13th January an aperture appeared on the upper edge of the cone from which at first a little lava issued forth, followed by the uprising of a cone which threw out projectiles accompanied by smoke, whilst the central crater continued to detonate more loudly and frequently. This little cone ultimately increased in size, until in April it filled the whole crater and rose four or five metres above the brim. At this time abundant lavas poured down from the base of the cone into the Atria del Cavallo, thence turned into the Fossa della Vetraria in the direction of the Observatory and towards the Crocella, where they accumulated to such an extent as to cover the hillside for a distance of about 300 metres; then turning below the Canteroni, formed a hillock without spreading much farther.

In October another small crater was formed by the falling in of the lava, which after a few days gave vent to smoke and several jets of lava; and towards the end of October the detonations increased, the smoke from the central crater issued forth more densely mixed with ashes, and the seismographical apparatus was much disturbed. On the 3rd and 4th November copious and splendid lava-streams coursed down the principal cone on its western side, but were soon exhausted; and in the beginning of 1872 the little cone, regaining vigour, began to discharge lava from the summit instead of the base as heretofore.

In the month of March 1873, with the full moon, the cone opened on the north-west side--the cleavage being indicated by a line of fumaroles--and lava issued from the base and poured down into the Atria as far as the precipices of Monte di Somma. On the 23rd April (another full moon) the activity of the craters increased, and on the evening of the 24th splendid lava-streams descended the cone in various directions, attracting on the same night the visits of a great many strangers. A lamentable event followed on the 26th. A party of visitors, accompanied by inexperienced guides, and contrary to the advice of Professor Palmieri, insisted on ascending to the place from which the lava issued. At half-past three on the morning of the 26th they were in the Atria del Cavallo, when the Vesuvian cone was rent in a north-west direction and a copious torrent of lava issued forth. Two large craters formed at the summit of the mountain, discharging incandescent projectiles and ashes. A cloud of smoke enveloped the unhappy visitors, who were under a hail-storm of burning projectiles. Eight were buried beneath it, or in the lava, while eleven were grievously injured.[11] The lava-stream, flowing over that of 1871 in the Atria, divided into two branches, the smaller one flowing towards Resina, but stopping before reaching the town; the larger precipitated itself into the Fossa della Vetraria, occupying the whole width of 800 metres, and traversing the entire length of 1300 metres in three hours. It dashed into the Fossa di Farone, and reached the villages of Massa and St. Sebastiano, covering a portion of the houses, and, continuing its course through an artificial foss, or trench, invaded cultivated ground and several villages. If it had not greatly slackened after midnight, from failure of supply at its source, it would have reached Naples by Ponticelli and flowed into the sea. The eruption towards the end of April had reached its height. The Observatory ridge was bounded on either side by two fiery streams, which rendered the heat intolerable. Simultaneously with the opening of the great fissure two large craters opened at the summit, discharging with a dreadful noise an immense cloud of smoke and ashes, with bombs which rose to a height of 1300 metres above the brim of the volcano.[12] The torrents of fire which threatened Resina, Bosco, and Torre Annunziata, and which devastated the fertile country of Novelle, Massa, St. Sebastiano, and Cerole, and two partially buried cities, the continual thunderings and growling of the craters, caused such terror, that numbers abandoned their dwellings, flying for refuge into Naples, while many Neapolitans went to Rome or other places. Fortunately, the paroxysm had now passed, the lava-streams stopped in their course, and the great torrent which passed the shoulders of the Observatory through the Fossa della Vetraria lowered the level of its surface below that of its sides, which appeared like two parallel ramparts above it. Had these streams continued to flow on the 27th of April as they had done on the previous night, they would have reached the sea, bringing destruction to the very walls of Naples. During this eruption Torre del Greco was upraised to the extent of two metres, and nearly all the houses were knocked down.

The igneous period of eruption having terminated, the ashes, lapilli, and projectiles became more abundant, accompanied by thunder and lightning. On the 28th they darkened the air, and the terrific noise of the mountain continuing or increasing, the terror at Resina, Portici, and Naples became universal. It seemed as though the tragic calamities of the eruption of A.D. 79 were about to be repeated. But gradually the force of the explosions decreased, and the noise from the crater became discontinuous, so that on the 30th the detonations were very few, and by the 1st May the eruption was completely over.

Such is a condensed account of one of the most formidable eruptions of our era. In the frontispiece of this volume a representation, taken (by permission) from a photograph by Negretti & Zambra, is given, showing the appearance of Vesuvius during the final stage of the eruption, when prodigious masses of smoke, steam, and illuminated gas issued forth from the summit and overspread the whole country around with a canopy which the light of the sun could scarcely penetrate.

It will be noticed in the above account that, concurrently with the full moon, there were two distinct and special outbreaks of activity; one occurring in March, the other in the month following. That the conditions of lunar and solar attraction should have a marked effect on a part of the earth's crust, while under the tension of eruptive forces, is only what might be expected. At full moon the earth is between the sun and the moon, and at new moon the moon is between the sun and the earth; under these conditions (the two bodies acting in concert) we have spring tides in the ocean, and a maximum of attraction on the mass of the earth. Hence the crust, which at the time referred to was under tremendous strain, only required the addition of that caused by the lunar and solar attractions to produce rupture in both cases, giving rise to increased activity, and the extrusion of lava and volatile matter. It may, in general, be safely affirmed that low barometric pressure on the one hand, and the occurrence of the syzygies (when the attractions of the sun and moon are in the same line) on the other, have had great influence in determining the crises of eruptions of volcanic mountains when in a state of unrest.

Contrast between the Northern and Southern Slopes.--Before leaving Vesuvius it may be observed that throughout all the eruptions of modern times the northern side of the mountain, that is the old crater and flank of

Somma, has been secure from the lava-flows, and has enjoyed an immunity which does not belong to the southern and western side. If we look at a map of the mountain showing the direction of the streams during the last three centuries,[13] we observe that all the streams of that period flowed down on the side overlooking the Bay of Naples; on the opposite side the wall of Monte di Somma presents an unbroken front to the lava-streams. From this it may be inferred that one side, the west, is weaker than the other; and consequently, when the lava and vapours are being forced upwards, under enormous pressure from beneath, the western side gives way under the strain, as in the case of the fissure of 1872, and the lava and vapours find means of escape. From what has happened in the past it is clear that no place on the western side of the mountain is entirely safe from devastation by floods of lava; while the prevalent winds tend to carry the ashes and lapilli, which are hurled into the air, in the same westerly direction.

[1] For an excellent view of this remarkable volcanic group see Judd's Volcanoes, 4th edition, p. 43.

[2] Plutarch, Life of Cassius; ed. Reiske, vol. iii. p. 240.

[3] Strabo gives the following account of the appearance and condition of Vesuvius in his day:--"Supra he loca situs est Vesuvius mons, agris cinctus optimis; dempto vertice, qui magna sui parte planus, totus sterilis est, adspectu sinereus, cavernasque ostendens fistularum plenas et lapidum colore fuliginoso, utpote ab igni exesorum. Ut conjectarum facere possis, ista loca quondam arsisse et crateras ignis habuisse, deinde materia deficiente restricta fuisse."--Rer. Geog., lib. v.

[4] A tablet over the entrance records this act of pious liberality, and is given by Phillips, loc. cit., p. 12.

[5] The stone pine, Pinus pinea, which Turner knew how to use with so much effect in his Italian landscapes.

[6] Bulwer Lytton's Last Days of Pompeii presents to the reader a graphic picture of the terrible event here referred to:--"The eyes of the crowd followed the gesture of the Egyptian, and beheld with ineffable dismay a vast vapour shooting from the summit of Vesuvius, in the form of a gigantic pine

tree; the trunk--blackness, the branches--fire! A fire that shifted and wavered in its hues with every moment--now fiercely luminous, now of a dull and dying red that again blazed terrifically forth with intolerable glare!... Then there arose on high the shrieks of women; the men stared at each other, but were speechless. At that moment they felt the earth shake beneath their feet; the walls of the theatre trembled; and beyond, in the distance, they heard the crash of falling roofs; an instant more and the mountain-cloud seemed to roll towards them, dark and rapid; at the same time it cast forth from its bosom a shower of ashes mixed with vast fragments of burning stone. Over the crushing vines--over the desolate streets--over the amphitheatre itself-- far and wide, with many a mighty splash in the agitated sea, fell that awful shower." A visit to the disinterred city will probably produce on the mind a still more lasting and vivid impression of the swift destruction which overtook this city.

[7] Quoted by Phillips, loc. cit., p. 45.

[8] Vesuvius, p. 72 et seq.

[9] Johnston-Lavis, "On the Geology of Monti Somma and Vesuvius," Quart. Jour. Geol. Soc., vol. 40 (1884).

[10] Palmieri, Eruption of Vesuvius in 1872, with notes, etc., by Robert Mallet, F.R.S. London, 1873.

[11] Those who lost their lives were medical students, and an Assistant Professor in the University, Antonio Giannone by name.

[12] Involving, as Mr. Mallet calculates, an initial velocity of projection of above 600 feet per second.

[13] Such as that given by Professor Phillips in his Vesuvius.

CHAPTER II.

ETNA.

(a.) Structure of the Mountain.--Etna, unlike Vesuvius, has ever been a burning mountain; hence it was well known as such to classic writers before the Christian era. The structure and features of this magnificent mountain have been abundantly illustrated by Elie de Beaumont,[1] Daubeny,[2] Baron von Waltershausen,[3] and Lyell,[4] of whose writings I shall freely avail myself in the following account, not having had the advantage of a personal examination of this region.

Structure of Etna.--So large is Etna that it would enclose within its ample skirts several cones of the size of Vesuvius. It rises to a height of nearly 11,000 feet above the waters of the Mediterranean,[5] and is planted on a floor consisting of stratified marine volcanic matter, with clays, sands, and limestones of newer Pliocene age. Its base is nearly circular, and has a circumference of 87 English miles. In ascending its flanks we pass successively over three well-defined physical zones: the lowest, or fertile zone, comprising the tract around the skirts of the mountain up to a level of about 2500 feet, being well cultivated and covered by dwellings surrounded by olive groves, fields, vineyards, and fruit-trees; the second, or forest zone, extending to a level of about 6270 feet, clothed with chestnut, oak, beech, and cork trees, giving place to pines; and the third, extending to the summit and called "the desert region," a waste of black lava and scori?with mighty crags and precipices, terminating in a snow-clad tableland, from which rises the central cone, 1100 feet high, emitting continually steam and sulphurous vapours, and in the course of almost every century sending forth streams of molten lava.

The forest zone is remarkable for the great number of minor craters which rise up from the midst of the foliage, and are themselves clothed with trees. Sartorius von Waltershausen has laid down on his map of Etna about 200 of these cones and craters, some of which, like those of Auvergne, have been broken down on one side. Many of these volcanoes of second or third magnitude lie outside the forest zone, both above and below it; such as the double hill of Monti Rossi, near Nicolosi, formed in 1659, which is 450 feet in height, and two miles in circumference at its base. Sir C. Lyell observes that these minor crater-cones present us with one of the most delightful and characteristic scenes in Europe. They occur of every variety of height and size, and are arranged in picturesque groups. However uniform they may appear when seen from the sea or the plains below, nothing can be more diversified than their shape when we look from above into their ruptured craters. The cones situated in the higher parts of the forest zone are chiefly clothed with lofty pines; while those at a lower elevation are adorned with chestnuts, oaks, and beech trees. These cones have from time to time been buried amidst fresh lava-streams descending from the great crater, and thus often become obliterated.

(b.) Val del Bove.--The most wonderful feature of Mount Etna is the celebrated Val del Bove (Valle del Bue), of which S. von Waltershausen has furnished a very beautiful plate[6]--a vast amphitheatre hewn out of the eastern flank of the mountain, just below the snow-mantled platform. It is a physical feature somewhat after the fashion of Monte Somma in Vesuvius, but exceeds it in magnitude as Etna exceeds Vesuvius. The Val del Bove is about five miles in diameter, bounded throughout three-fourths of its circumference by precipitous walls of ashes, scori? and lava, traversed by innumerable dykes, and rising inwards to a height of between 3000 and 4000 feet. Towards the east the cliffs gradually fall to a height of about 500 feet, and at this side the vast chasm opens out upon the slope of the mountain. At the head of the Val del Bove rises the platform, surmounted by the great cone and crater. It will thus be seen that by means of this hollow we have access almost to the very heart of the mountain.

What is very remarkable about the structure of this valley is that the beds exhibit "the ququ versal dip"--in other words, they dip away on all sides from the centre--which has led to the conclusion that in the centre is a focus of eruption which had become closed up antecedently to the formation of the valley itself. Lyell has explained this point very clearly by showing that this focus had ceased to eject matter at some distant period, and that the existing crater at the summit of the mountain had poured out its lavas over those of the extinct orifice. This was prior to the formation of the Val del Bove itself; and the question remains for consideration how this vast natural amphitheatre came to be hollowed out; for its structure shows unquestionably that it owes its form to some process of excavation.

In the first place, it is certainly not the work of running water, as in the case of the ca 駒 ns of Colorado; the porous matter of which the mountain is formed is quite incapable of originating and supporting a stream of sufficient volume to excavate and carry away such enormous masses of matter within the period required for the purpose. We must therefore have recourse to some other agency. Numerous illustrations are to be found of the explosive action of volcanoes in blowing off either the summits of mountains, or portions of their sides. For example, there is reason for believing that the first result of the renewed energy of Vesuvius was to blow into the air the upper surface of the mountain. Again, so late as 1822, during a violent earthquake in Java, a country which has been repeatedly devastated by earthquakes and volcanic eruptions, the mountain of Galongoon, which was covered by a dense forest, and situated in a fertile and thickly-peopled region, and had never within the period of tradition been in activity, was thus ruptured by internal forces. In the month of July 1822, after a terrible earthquake, an explosion was heard, and immense columns of boiling water, mixed with mud and stones, were projected from the mountain like a water-spout, and in falling filled up the valleys, and covered the country with a thick deposit for many miles, burying villages and their inhabitants. During a subsequent eruption great blocks of basalt were thrown to a distance of seven miles; the result of all being that an enormous semicircular gulf was formed between the summit and the plain, bounded by steep cliffs, and bearing considerable resemblance to the Val del Bove. Other examples of the power of volcanic explosions might be cited; but the above are sufficient to show that great hollows may thus be formed either on the summits or flanks of volcanic mountains. Chasms may also be formed by the falling in of the solidified crust, owing to the extrusion of molten matter from some neighbouring vent of eruption; and it is conceivable that by one or other of these processes the vast chasm of the Val del Bove on the flanks of Etna may have been produced.

(c.) The Physical History of Etna.--The physical history of Etna seems to be somewhat as follows:--

First Stage.--Somewhere towards the close of the Tertiary period--perhaps early Pliocene or late Miocene--a vent of eruption opened on the floor of the Mediterranean Sea, from which sheets of lava were poured forth, and ashes mingled with clays and sands, brought down from the neighbouring lands, were strewn over the sea-bed. During a pause in volcanic activity, beds of limestone with marine shells were deposited.

Second Stage.--This sea-bed was gradually upraised into the air, while fresh sheets of lava and other ejecta were accumulated round the vents of eruption, of which there were two principal ones--the older under the present Val del Bove, the newer under the summit of the principal cone. Thus was the mountain gradually piled up.

Third Stage.--The vent under the Val del Bove ceased to extrude more matter, and became extinct. Meanwhile the second vent continued active, and, piling up more and more matter round the central crater, surmounted the former vent, and covered its ejecta with newer sheets of lava, ashes, and lapilli, while numerous smaller vents, scattered all over the sides of the mountain, gave rise to smaller cones and craters.

Fourth Stage.--This stage is signalised by the formation of the Val del Bove through some grand explosion, or series of explosions, by which this vast chasm was opened in the side of the mountain, as already explained.

Fifth Stage.--This represents the present condition of the mountain, whose height above the sea is due, not only to accumulation of volcanic materials round the central cone, but to elevation of the whole island, as evinced by numerous raised beaches of gravel and sand, containing shells and other forms of marine species now living in the waters of the Mediterranean.[7] Since then the condition and form of the mountain has remained very much the same, varied only by the results of occasional eruptions.

(d.) Dissimilarity in the Constitution of the Lavas of Etna and Vesuvius.-- Before leaving the subject we have been considering, it is necessary that I should mention one remarkable fact connected with the origin of the lavas of Etna and Vesuvius respectively; I refer to their essential differences in mineral composition. It might at first sight have been supposed that the lavas of these two volcanic mountains--situated at such a short distance from each other, and evidently along the same line of fracture in the crust--would be of the same general composition; but such is not the case. In the lava of Vesuvius leucite is an essential, and perhaps the most abundant mineral. It is called by Zirkel Sanidin-Leucitgestein. (See Plate IV.) But in that of Etna this mineral is (as far as I am aware) altogether absent. We have fortunately abundant means of comparison, as the lavas of these two mountains have been submitted to close examination by petrologists. In the case of the Vesuvian lavas, an elaborate series of chemical analyses and microscopical observations have been made by the Rev. Professor Haughton, of Dublin University, and the author,[8] from specimens collected by Professor Guiscardi from the lava-flows extending from 1631 to 1868, in every one of which leucite occurs, generally as the most abundant mineral, always as an essential constituent. On the other hand, the composition of the lavas of Etna, determined by Professor A. von Lasaulx, from specimens taken from the oldest sheets of lava down to those of the present day, indicates a rock of remarkable uniformity of composition, in which the components are plagioclase felspar, augite, olivine, magnetite, and sometimes apatite; but of leucite we have no trace.[9] In fact, the lavas of Etna are very much the same in composition as the ordinary basalts of the British Isles, while those of Vesuvius are of a different type. This seems to suggest an origin of the two sets of lavas from a different deep-seated magma; the presence of leucite in such large quantity requiring a magma in which soda is in excess, as compared with that from which the lavas of Etna have been derived.[10]

[1] Memoires pour Servir, etc., vol. ii.

[2] Daubeny, Volcanoes, p. 270.

[3] Von Waltershausen, Der Etna, edited by A. von Lasaulx.

[4] Lyell, Principles of Geology, vol. ii., edition 1872.

[5] Its height, as determined by Captain Smyth in 1875 trigonometrically, was 10,874 feet, and afterwards by Sir J. Herschel barometrically, 10,872 feet.

[6] Atlas des Etna (Weimar, 1858), in which the different lava-streams of 1688, 1802, 1809, 1811, 1819, 1824, and 1838 are delineated.

[7] Sir William Hamilton observes that history is silent regarding the first eruptions of Etna. It was in activity before the Trojan War, and even before the arrival of the "Sizilien" settlers. Diodorus and Thucydides notice the earliest recorded eruptions, those from 772 to 388 B.C., during which time the mountain was thrice in eruption. Later eruptions took place in the year 140, 135, 125, 122 B.C. In the year 44 B.C., in the reign of Julius Caesar, there was a very violent outburst of volcanic activity.--Neuere Beobachtungen die Vulkane Italiens und am Rhein, p. 173, Frankfurt (1784).

[8] "Report on the Chemical and Mineralogical Characters of the Lavas of Vesuvius from 1631 to 1868," Transactions of the Royal Irish Academy, vol.

xxvi. (1876). In the lava of 1848 leucite was found to reach 44.9 per cent. of the whole mass. In that of Granatello, 1631, it reaches its lowest proportion-- viz., 3.37 per cent.

[9] A. von Lasaulx, in Von Waltershausen's Der na, Book II., x. 423.

[10] The view of Professor Judd, that leucite easily changes into felspar, and that some ancient igneous rocks which now contain felspar were originally leucitic, does not seem to be borne out by the above facts. In such cases the felspar crystals ought to retain the forms of leucite. See Volcanoes, 4th edition, p. 268.

CHAPTER III.

THE LIPARI ISLANDS, STROMBOLI.

(a.) A brief account of this remarkable group of volcanic islands must here be given, inasmuch as they seem to be representatives of a stage of volcanic action in which the igneous forces are gradually losing their energy. According to Daubeny, the volcanic action in these islands seems to be developed along two lines, nearly at right angles to each other, one parallel to that of the Apennines, beginning with Stromboli, intersecting Panaria, Lipari, and Vulcano; the other extending from Panaria to Salina, Alicudi, and Felicudi, and again visible in the volcanic products which make their appearance at Ustica. (See Map, Fig. 11.) The islands lie between the north coast of Sicily and that of Italy, and from their position seem to connect Etna with Vesuvius; but this is very problematical, as would appear from the difference of their lavas. The principal islands are those of Stromboli, Panaria, Lipari, Vulcano, Salina, Felicudi, and Alicudi. These three last are extinct or dormant, but Salina contains a crater, rising, according to Daubeny, not less than 3500 feet above the sea.[1] Vulcano (referred to by Strabo under the name of Hiera) consists of a crater which constantly emits large quantities of sulphurous vapours, but was in a state of activity in the year 1786, when, after frequent earthquake shocks and subterranean noises, it vomited forth during fifteen days showers of sand, together with clouds of smoke and flame, altering materially the shape of the crater from which they proceeded.

The islands of Lipari are formed of beds of tuff, penetrated by numerous dykes of lava, from which uprise two or three craters, formed of pumice and obsidian passing into trachyte. Volcanic operations might have here been said to be extinct, were it not that their continuance is manifested by the existence of hot springs and "stufes," or vapour baths, at St. Calogero, about four miles from the town of Lipari. Daubeny considers it not improbable that this island may have had an active volcano even within the historical period, a view which is borne out by the statement of Strabo.[2]

(b.) But by far the most remarkable island of the group, as regards its present volcanic condition, is Stromboli, which has ever been in active eruption from the commencement of history down to the present day. Professor Judd, who visited this island in 1874, and has produced a striking representation of its aspect,[3] gives an account of which I shall here avail myself.[4] The island is of rudely circular outline, and rises into a cone, the summit of which is 3090 feet above the level of the Mediterranean. From a point on the side of the mountain masses of vapour are seen to issue, and these unite to form a cloud over the summit; the outline of this vapour-cloud varying continually according to the hygrometric state of the atmosphere, and the direction and force of the wind. At the time of Professor Judd's visit, the vapour-cloud was spread in a great horizontal stratum overshadowing the whole island; but it was clearly seen to be made up of a number of globular masses, each of which is a product of a distinct outburst of volcanic forces. Viewed at night-time, Stromboli presents a far more striking and singular spectacle. When watched from the deck of a vessel, a glow of red light is seen to make its appearance from time to time above the summit of the mountain; it may be observed to increase gradually in intensity, and then as gradually to die away. After a short interval the same appearances are repeated, and this goes on till the increasing light of dawn causes the phenomenon to be no longer visible. The resemblance presented by Stromboli to a "flashing light" on a most gigantic scale is very striking, and the mountain has long been known as "the lighthouse of the Mediterranean."

The mountain is built up of ashes, slag, and scori? to a height of (as already stated) over 3000 feet above the surface of the sea; but, as Professor Judd observes, this by no means gives a just idea of its vast bulk. Soundings in the sea surrounding the island show that the bottom gradually shelves around the shores to a depth of nearly 600 fathoms, so that Stromboli is a great conical mass of cinders and slaggy materials, having a height above its floor of about 6600 feet, and a base the diameter of which exceeds four miles.

The crater of Stromboli is situated, not at the apex of the cone, but at a distance of 1000 feet below it. The explosions of steam, accompanied by the roaring as of a smelting furnace, or of a railway engine when blowing off its steam, are said by Judd to take place at very irregular intervals of time, "varying from less than one minute to twenty minutes, or even more." On the other hand, Hoffmann describes them as occurring at "perfectly regular intervals," so that, perhaps, some variation has taken place within the interval of about forty years between each observation. Both observers agree in stating that lava is to be seen welling up from some of the apertures within the crater, and pouring down the slope towards the sea, which it seldom or never reaches.[5] The intermittent character of these eruptions appears to be due, as Mr. Scrope has suggested, to the exact proportion between the expansive and repressive forces; the expansive force arising from the generation of a certain amount of aqueous vapour and of elastic gas; the repressive, from the pressure of the atmosphere and from the weight of the superincumbent volcanic products. Steam is here, as in a steam-engine, not the originating agent in the phenomena recorded; but the result of water coming in contact with molten lava constantly welling up from the interior, by which it is converted into steam, which from time to time acquires sufficient elastic force to produce the eruptions; the water being obviously derived from the surrounding sea, which finds its way by filtration through fissures, or through the porous mass of which the mountain is formed. Were it not for the access of water this volcano would probably appear as a fissure-cone extruding a small and continuous stream of molten lava. The adventitious access of the sea water gives rise to the phenomena of intermittent explosions. The vitality of the volcano is therefore due, not to the presence of water, but to the welling up of matter from the internal reservoir through the throat of the volcano.

Pantelleria.--This island, lying between the coast of Sicily and Cape Bon in Africa, is wholly volcanic. It has a circumference of thirty miles, and from its centre rises an extinct crater-cone to a height of about 3000 feet. The flanks of this volcano are diversified by several fresh craters and lava-streams, while hot springs burst out with a hissing noise on its southern flank, showing that molten matter lies below at no very great depth.

This island probably lies along the dividing line between the non-volcanic and volcanic region of the Mediterranean, and is consequently liable to intermittent eruptions. It was at a short distance from this island that the remarkable submarine outburst of volcanic forces took place on October 17th, 1891, for an account of which we are indebted to Colonel J. C. Mackowen.[6] On that day, after a succession of earthquake shocks, the inhabitants were startled by observing a column of "smoke" rising out of the sea at a distance of three miles, in a north-westerly direction. The Governor, Francesco Valenza, having manned a boat, rowed out towards the fiery column, and on arriving found it to consist of black scoriaceous bombs, which were being hurled into the air to a height of nearly thirty yards; some of them burst in the air, others, discharging steam, ran hissing over the water; many of them were very hot, some even red-hot. One of these bombs, measuring two feet in diameter, was captured and brought to shore. It was observed that after the eruption the earthquake shocks ceased. A vast amount of material was cast out of the submarine crater, forming an island 500 yards in length and rising up to nine feet above the surface, but after a few days it was broken up and dispersed over the sea-bed by the action of the waves.

[1] Volcanoes, p. 262. These islands are described by Hoffmann, Poggendorf Annal., vol. xxvi. (1832); also by Lyell, Principles of Geology, vol. ii., and by Judd, who personally visited them, and gives a very vivid account of their appearance and structure.

[2] Strabo, lib. vi.

[3] Judd, Volcanoes, p. 8.

[4] Stromboli has also been described by Spallanzani, Hoffmann, Daubeny, and others. The account of Judd is the most recent. Of this island Strabo says, "Strongyle a rotundate figur?sic dicta, ignita ipsa quoque, violentia flammarum minor, fulgore excellens; ibi habitasse olum ajunt."--Lib. vi.

[5] Poggend. Annal., vol. xxvi., quoted by Daubeny.

[6] Communicated by Captain Petrie to the Victoria Institute, 1st February 1892. See also a detailed and illustrated account of the eruption communicated by A. Ricco to the Annali dell' Ufficio centrale Meteorologico e

Geodonamico, Ser. ii., Parte 3, vol. xi. Summarised by Mr. Butler in Nature, April 21, 1892.

CHAPTER IV.

THE SANTORIN GROUP.

a.) Before leaving the subject of European active volcanoes, it is necessary to give some account of the remarkable volcanic island of Santorin, in the Grecian archipelago. This island for 2000 years has been the scene of active volcanic operations, and in its outline and configuration, both below and above the surface of the Mediterranean, presents the aspect of a partially submerged volcanic mountain. (See Section, Fig. 13.) If, for example, we can imagine the waters of the sea to rise around the flanks of Vesuvius until they have entered and overflowed to some depth the interior caldron of Somma, thus converting the old crater into a crescent-shaped island, and the cone of Vesuvius into an island--or group of islands--within the caldron, then we shall form some idea of the appearance and structure of the Santorin group.

Form of the Group.--The principal island, Thera, has somewhat the shape of a crescent, breaking off in a precipitous cliff on the inner side, but on the outer side sloping at an angle of about fifteen degrees into deep water. Continuing the curvature of the crescent, but separated by a channel, is the island of Therasia; and between this and the southern promontory of Thera is another island called Aspronisi. All these islands, if united, would form the rim of a crater, in which the volcanic matter slopes outward into deep water, descending at a short distance to a depth of 200 fathoms and upwards. In the centre of the gulf thus formed rise three islands, called the Old, New, and Little Kaimenis. These may be regarded as cones of eruption, which history records as having been thrown up at successive intervals. According to Pliny, the year 186 B.C. gave birth to Old Kaimeni, also called Hiera, or the Sacred Isle; and in the first year of our era Thera (the Divine) made its appearance above the water, and was soon joined to the older island by subsequent eruptions. Old Kaimeni also increased in size by the eruptions of 726 and 1427. A century and a half later, in 1573, another eruption produced the cone and crater called Micra-Kaimeni. Thus were formed, or rather were rendered visible above the water, the central craters of eruption; and between these and the inner cliff of Thera and Therasia is a ring of deep water, descending to a depth of over 200 fathoms. So that, were these islands raised out of the sea, we should have presented to our view a magnificent circular crater about six miles in diameter, bounded by nearly vertical walls of rock from 1000 to 1500 feet in height, and ruptured at one point, from the centre of which would rise two volcanic cones--namely, the Kaimenis--one with a double crater, still foci of eruption, and from time to time bursting forth in paroxysms of volcanic energy, of which those of 1650, 1707, and 1866 were the most violent and destructive.[1] Of this last I give a bird's-eye view (Fig. 14).

The only rock of non-volcanic origin in these islands consists of granular limestone and clay slate forming the ridge of Mount St. Elias, which rises to a height of 1887 feet at the south-eastern side of the island of Thera, crossing the island from its outer margin nearly to the interior cliff, so that the volcanic materials have been piled up along its sides. The rocks of St. Elias are much more ancient than any of the volcanic materials around; and, as Bory St. Vincent has shown, have been subjected to the same flexures, dip and strike, as those sedimentary rocks which go to form the non-volcanic islands of the Grecian archipelago.

(b.) Origin of the Santorin Group.--In reference to the origin of the Santorin group, Lyell regards it as a remnant of a great volcanic mountain which possessed a focus of eruption rising in the position of the present foci, but afterwards partially destroyed and the whole submerged to a depth of over 1000 feet. But another explanation is open to us, and one not inconsistent with what we now know of the physical changes to which the Mediterranean has been subjected since early Tertiary times. To my mind it is difficult to conceive how such a volcanic mountain as that of Santorin could have been formed under water; while, on the other hand, its physical structure and contour bear so striking a resemblance (as already observed) to those of Vesuvius and Rocca Monfina that we are much tempted to infer that it had a somewhat similar origin. Now we know that Vesuvius was built up by means of successive eruptions taking place under the air; and the question arises whether it could be possible that Santorin had a similar origin owing to the waters of the Mediterranean having been temporally lowered at a later Tertiary epoch. It has been stated by M. Fouqu?that the age of the more ancient volcanic beds of Santorin belong, as shown by the included fossils, to the newer Pliocene epoch. These are of course the unsubmerged, and therefore more recent strata, and may have been recently upheaved during one or more of the outbursts of volcanic energy. But it seems an impossibility that the Gulf of Santorin, with its precipitous walls and deep circular interior channel, as shown by the Ideal Section (Fig. 13), could have been formed otherwise than under the air. We are led, therefore, to inquire whether there was a time in the history of the Mediterranean, since the Eocene period, when the waters were lower than at present. That this was the case we have clear evidence. The remains of elephants, hippopotami, and other animals, which have been discovered in great numbers in the Maltese caves, show that this island was united to Sicily, and this again to Europe, during the later Pliocene epoch, so as to have become the abode of an Europasian fauna. According to Dr. Wallace, a causeway of dry land existed, stretching from Italy to Tunis in North Africa through the Maltese Islands--an inference involving the lowering of the waters of the Mediterranean by several hundred feet.[2] There is every reason for supposing that the old volcano of Santorin was in active eruption at this period; and its history may be considered to be similar to that of Vesuvius until, at the rising of the waters during the Pluvial (or Post-Pliocene) epoch, during which they rose higher than at present, Santorin was converted into a group of islands, slightly differing in form from those of the present day. This view seems to meet the difficulties regarding the origin of this group, difficulties which Lyell had long since clearly recognised.

(c.) Limit of the Mediterranean Volcanic Region.--With the Santorin group we conclude our account of the active European volcanoes. It may be observed, however, that from some cause not ascertained the volcanic districts of the Mediterranean and its shores are confined to the north side of that great inland sea; so that as regards vulcanicity the African coast presents a striking contrast to that of the opposite side. If we draw a line from the shores of the Levant to the Straits of Gibraltar, by Candia, Malta, and to the south of Pantelleria and Sardinia, we shall find that the volcanic islands and districts of the mainland lie to the north of it.[3] This has doubtless some connection with the internal geological structure. The immunity of the Libyan desert from volcanic irruptions is in keeping with the remarkably undisturbed condition of the Secondary strata, which seldom depart much from the horizontal position; while the igneous rocks of the Atlas mountains are probably of great geological antiquity. On the other hand, the Secondary and Tertiary formations of the northern shores and islands of the Mediterranean are generally characterised by the highly-inclined, flexured, and folded position of the strata. Hence we may suppose that the crust over the region lying to the north of the volcanic line, owing to its broken and ruptured condition, was less able to resist the pressure of the internal forces of eruption than that lying to the south of it; and that, in consequence, vents and fissures of eruption were established over the former of these regions, while they are absent in the latter.

[1] Fuller details will be found in Daubeny's Volcanoes, chap. xviii., and Lyell's Principles of Geology, vol. ii. p. 65 (edition 1872). The bird's-eye view is taken from this latter work by kind permission of the publisher, Mr. J. Murray, as also the accompanying Ideal Section, Fig. 13.

[2] Wallace, Geographical Distribution of Animals (1876). The author's Sketch of Geological History, p. 130 (Deacon & Co., 1887).

[3] The volcanic area lying to the north of this line will include Sardinia, Sicily, Pantelleria, the Grecian Archipelago, Asia Minor, and Syria; the non-volcanic area lying to the south of this line will include the African coast, Malta, Isles of Crete and Cyprus. The Isle of Pantelleria is apparently just on the line, which, continued eastward, probably follows the north coast of Cyprus, parallel to the strike of the strata and of the central axis of that island.--See "Carte G 鬧 logique de l'e de Chypre, par MM. Albert Gaudry et Amed Damour" (1860).

CHAPTER V.

EUROPEAN EXTINCT OR DORMANT VOLCANOES.

We are naturally led on from a consideration of the active volcanoes of Europe to that of volcanoes which are either dormant or extinct in the same region. Such are to be found in Italy, Central France, both banks of the Rhine and Moselle, the Westerwald, Vogelsgebirge, and other districts of Germany; in Hungary, Styria, and the borders of the Grecian archipelago. But the subject is too large to be treated here in detail; and I propose to confine my observations to some selected cases which are to be found in Southern Italy, Central France, and the Rhenish districts, where the volcanic features are of so recent an age as to preserve their outward form and structure almost intact.

(a.) Southern Italy.--Extinct volcanoes and volcanic rocks occupy considerable tracts between the western flanks of the Apennines and the Mediterranean coast in the Neapolitan and Roman States, forming the remarkable group of the Phlegren fields, with the adjoining islands of Ischia, Procida, Nisida, Vandolena, Ponza, and Palmarola; at Melfi and Avellino. All the region around Rome extending along the western slopes of the Apennines from Velletri to Orvieto, together with Mount Annato in Tuscany, is formed of volcanic material, and the same may be said of a large part of the island of Sardinia. From these districts I shall select some points which seem to be of special interest.

Monte Nuovo and the Phlegren Fields.--The tract of which this celebrated district forms a part lies as it were in a bay of the Apennine limestone of Jurassic age. The floor of this bay is composed of puzzolana, a name given to beds of volcanic tuff of great thickness, and rising into considerable hills in the vicinity of the city of Naples, such as that of St. Elmo. Its composition is peculiar, as it is chiefly formed of small pieces of pumice, obsidian, and trachyte, in beds alternating with loam, ferriferous sand, and fragments of limestone. It is evidently of marine formation, as Sir William Hamilton, Professor Pilla, and others have detected sea-shells therein, of the genera Ostr, Cardium, Pecten and Pectunculus, Buccinum, etc. It is generally of a greyish colour, and sometimes sufficiently firm to be used as a building stone. The Roman Campagna is largely formed of similar materials, which were deposited at a time when the districts in question were submerged, and matter was being erupted from volcanic vents at various points around, and spread over the sea-bed.

Such is the character of the general floor on which the more recent crater-cones of this district have been built. These are numerous, and all extinct with the exception of the Solfatara, near Puzzuoli, from which gases mixed with aqueous vapour are continually being exhaled. The gases consist of sulphuretted hydrogen mixed with a minute quantity of muriatic acid.[1] This district is also remarkable for containing several lakes occupying the interiors of extinct craters; amongst others, Lake Avernus, which, owing to its surface having been darkened by forests, and in consequence of the effluvia arising from its stagnant waters, has had imparted to it a character of gloom and terror, so that Homer in the Odyssey makes it the entrance to hell, and describes the visit of Ulysses to it. Virgil follows in his steps. Another lake of similar origin is Lake Agnano. Here also is the Grotto del Cane, a cavern from which are constantly issuing volumes of carbonic acid gas combined with much aqueous vapour, which is condensed by the coldness of the external air, thus proving the high temperature of the ground from which the gaseous vapour issues. This whole volcanic region, so replete with objects of interest,[2] may be considered, as regards its volcanic character, in a moribund condition; but that it is still capable of spasmodic movement is evinced by the origin of Monte Nuovo, the most recent of the crater-cones of the district. This mountain, rising from the shore of the Bay of Bai? was suddenly formed in September 29th, 1538, and rises to a height of 440 feet above the sea-level. It is a crater-cone, and the depth of the crater has been determined by the Italian mineralogist Pini to be 421 English feet; its bottom is thus only 19 feet above the sea-level. A portion of the base of the cone is considered partly to occupy the site of the Lucrine Lake, which was itself nothing more than the crater of a pre-existent volcano, and was almost entirely filled up during the explosion of 1538. Monte Nuovo is composed of ashes, lapilli, and pumice-stones; and its sudden formation, heralded by earthquakes, and accompanied by the ejection of volcanic matter mixed with fire and water, is recorded by Falconi, who vividly depicts the terror and consternation of the inhabitants of the surrounding country produced by this sudden and terrible outburst of volcanic forces.[3]

(b.) Central Italy and the Roman States.--The tract bordering the western slopes of the Apennines northward from Naples into Tuscany, and including the Roman States, is characterised by volcanic rocks and physical features of remarkable interest and variety. These occur in the form of extinct craters, sometimes filled with water, and thus converted into circular lakes; or of extensive sheets and conical hills of tuff; or, finally, of old necks and masses of trachyte and basalt, sometimes exhibiting the columnar structure. The Eternal City itself is built on hills of volcanic material which some observers have supposed to be the crater of a great volcano; but Ponzi, Brocchi, and Daubeny all concur in the opinion that this is not the case, as will clearly appear from the following account.

The geological structure of the valley of the Tiber at Rome is very clearly described by Professor Ponzi in a memoir published in 1850, from which the accompanying section is taken.[4] (Fig. 16.) From this it will be seen that "the Seven-hilled City" is built upon promontories of stratified volcanic tuff, of which the Campagna is formed, breaking off along the banks of the Tiber, the hills being the result of the erosion, or denudation, of the strata along the side of the river valley. As the strata dip from west to east across the course of the river, it follows that those on the western banks are below those on the opposite side; and thus the marine sands and marls which underlie the volcanic tuff, and are concealed by it along the eastern side of the valley, emerge on the west, and form the range of hills on that side. Such being the structure of the formations under Rome, it is evident that it is not "built on a volcano."

The tuff contains fragments of lava and pebbles of Apennine limestone, and was deposited under the waters of an extensive lake at a time when volcanic action was rife amidst the Alban Hills. This lacustrine formation rests in turn on deposits of marine origin, containing oysters, patella and other sea-shells, of which the chain of hills on the right bank of the Tiber is chiefly formed.

The district around Albano lying to the south of Rome is of peculiar interest from the assemblage of old crater-lakes which it contains; as, for instance, those of Albano, Vallariccia, Nemi, Juturna, and the lake of Gabii. The lake of Albano, one of the most beautiful sheets of water in the world, is about six miles in circumference, and surrounded by beds of peperino, a variety of tuff presenting a bright, undecomposed aspect when newly broken. The level of this lake was lowered by the Romans during the siege of Veii by means of a tunnel, so that the waters are 200 feet lower than the level at which they originally stood. In the same district is the lake of Nemi, very regular in its circular outline; that of Juturna lying near the foot of the Alban Hills, and that of Ariccia lying in a deep hollow eight miles in circumference;--all may be supposed to have been the craters of extinct volcanoes, both by reason of their shape and of the materials of which they are formed. All these old craters are, however, according to Daubeny, "only the dependencies and offshoots, as it were, of the great extinct volcano, the traces of which still remain upon the summit of the Alban Hills, and which is comparable in its form to that of Vesuvius, as it is surrounded by an outer circle of volcanic rock comparable to that of Somma."[5]

To the north of the city of Rome are several crateriform lakes, some of which are of great size, such as that of Bolsena, over twenty miles in circumference, and the Lago di Bracciano, almost as large, and lying about twelve miles from the city. These extensive sheets of water are surrounded by banks of tuff and volcanic sand, in which fragments of augite, leucite, and crystals of titanite are distributed. The town of Viterbo is built up at the foot of a steep hill called Monte Cimini, the lower part of which is composed of trachyte; this is surmounted by tuff, which appears to have been ejected from an extinct crater occupying the summit of the mountain, and now converted into a lake called the Lake of Vico. This crater is perfectly circular, and from its centre rises a little conical hill covered by trees.

(c.) Physical History.--Space does not permit of a fuller description of the remarkable volcanic features of the tract lying along the western slope of the Apennines; but from what has been stated it will be clear that volcanic forces have been in operation at one time on a grand scale in the Roman States and the South of Tuscany, over a tract extending from Mount Annato to Velletri and Segni.

This tract was separated from that of the Neapolitan volcanic region by a range of limestone hills of Jurassic age between Segni and Gaeta, a protrusion of the Alban Hills westward; but the general structure and physical history of both regions are probably very similar, with the exception that the igneous forces still retain their vitality in the more southerly region. In the case of the Roman volcanic district, a bay seems to have been formed about the close of the Miocene period, bounded on all sides but the west by hills of limestone, over whose bed strata of marl, sandstone, and conglomerate were deposited. This tract was converted by subsequent movements into a fresh-water lake, and contemporaneously volcanic operations commenced over the whole region, and beds of tuff, often containing blocks of rock ejected from neighbouring craters, were deposited over those of marine origin. Meanwhile numerous crater-cones were thrown up; and, as the land gradually rose, the waters of the lake were drained off, leaving dry the Campagna and plain of the Tiber. Ultimately the volcanic fires smouldered down and died out, whether within the historic epoch or not is uncertain; lakes were formed within the now dormant craters, and the face of nature gradually assumed a more placid and less forbidding aspect over this memorable region, destined to be the site of Rome, the Mistress of the World.

[1] As determined by Daubeny in 1825.

[2] Including the ruins of the Temple of Serapis, whose pillars are perforated by marine boring shells up to a height of about 16 feet from their base; indicating that the land had sunk down beneath the sea, and afterwards been elevated to its present level.

[3] The account of Falconi, and another by Pietro Giacomo di Toledo, are given by Sir W. Hamilton, op. cit., p. 198, and also reproduced by Sir C. Lyell, Principles, vol. i. p. 608.

[4] Guiseppe Ponzi, "Sulla storia fisica del Bacino di Roma," Annali di Scienze Fisiche (Roma, 1850).

[5] Daubeny, Volcanoes, p. 171.

CHAPTER VI.

EXTINCT VOLCANOES OF CENTRAL FRANCE.

(a.) General Structure of the Auvergne District.--From a granitic and gneissose platform situated near the centre of France, and separated from the western spurs of the Alps by the wide valley of the Rhone, there rises a group of volcanic mountains surpassing in variety of form and structure any similar mountain group in Europe, and belonging to an epoch ranging from the Middle Tertiary down almost to the present day. This volcanic group of mountains gives rise to several important rivers, such as the Loire, the Allier, the Soule (a branch of the Loire), the Creuse, the Dordogne, and the Lot; and in the Plomb du Cantal attains an elevation of 6130 feet above the sea. Its southern section, that of Mont Dore, the Cantal, and the Haute Loire, is characterised by magnificent valleys, traversing plateaux of volcanic lava, and exhibiting the results of river erosion on a grand scale; while its northern section, that of the Puy de De, presents to us a varied succession of volcanic crater-cones and domes, with their extruded lava-streams, almost as fresh and unchanged in form as if they had only yesterday become extinct. A somewhat similar, but less important, chain of extinct volcanoes also occurs in the Velay and Vivarais, between the upper waters of the Loire and the Allier, in the vicinity of the town of Le Puy.[1] The principal city in this region is Clermont-Ferrand, lying near the base of the Puy de De, and ever memorable as the birthplace of Blaise Pascal.[2]

The physical structure of this region is on the whole very simple. The fundamental rocks consist of granite and gneiss passing into schist, all of extreme geological antiquity, forming a vast platform gradually rising in a southerly direction towards the head waters of the Loire and the Allier in the Departments of Haute Loire, Loze, and Ardhe. On this platform are planted the whole of the volcanic mountains. (See Fig. 17.)

The granitic plateau is bounded on the east, throughout a distance of about 50 miles, by the wide and fertile plain of Clermont, watered by the Allier and its numerous branches descending from the volcanic mountains, and is about 25 miles in width from east to west in the parallel of Clermont, but gradually narrowing in a southerly direction, till at Brioude it becomes an ordinary mountain ravine. The eastern margin of the plain is formed by another granitic ridge expanding into a plateau towards the south, and joining in with that already described; but towards the north and directly east of Clermont it forms a high ridge traversed by the railway to St.ienne and Lyons, and descending towards the east into the valley of the Loire. No more impressive view is to be obtained of the volcanic region than that from the summit of this second ridge, on arriving there towards evening from the city of Lyons. At your feet lies the richly-cultivated plain of Clermont, dotted with towns, villages, and hamlets, and decorated with pastures, orchards, vineyards, and numerous trees; while beyond rises the granitic plateau, breaking off abruptly along the margin of the plain, and deeply indented by the valleys and gorges along which the streams descend to join the Allier. But the chief point of interest is the chain of volcanic crater-cones and dome-shaped eminences which rise from the plateau, amongst which the Puy de De towers supreme. Their individual forms stand out in clear and sharp relief against the western sky, and gradually fade away towards the south into the serried masses of Mont Dore and Cantal, around whose summits the evening mists are gathering. Except the first view of the Mont Blanc range from the crest of the Jura, there is no scene perhaps which is calculated to impress itself more vividly on the memory than that here faintly described.[3]

(b.) The Vale of Clermont.--The plain upon which we look down was once the floor of an extensive lake, for it is composed of various strata of sand, clay, marl, and limestone, containing various genera and species of fresh-water shells. These strata are of great thickness, perhaps a thousand feet in some places; and along with such shells as Paludina, Planorbis, and Lim are also found remains of various other animals, such as fish, serpents, batrachians, crocodiles, ruminants, and those of huge pachyderms, as Rhinoceros, Dinotherium, and Cotherium. This great lake, occupying a hollow in the old granitic platform of Central France, must have been in existence for an extensive period, which MM. Pomel, Aymard, and Lyell all unite in referring to that of the Lower Miocene. But what is to us of special interest is the fact that, in the deposits of this lake of the Haute Loire, with the exception of the very latest, there is no intermixture of volcanic products such as might have been expected to occur if the neighbouring volcanoes had been in activity during its existence. Hence it may be supposed that, as Scrope suggested, the waters of the lake were drained off owing to the disturbance in the levels of the country caused by the first explosions of the Auvergne volcanoes.[4] If this be so, then we possess a key by which to determine the period of the first formation of volcanoes in Central France; for, as the animal remains enclosed in the lacustrine deposits of the Vale of Clermont belong to the early Miocene stage, and the earliest traces of contemporaneous volcanic ejecta are found only in the uppermost deposits, we may conclude that the first outburst of volcanic action occurred towards the close of the Miocene period--a period remarkable for similar exhibitions of internal igneous action in other parts of the world.

(c.) Successive Stages of Volcanic Action in Auvergne.--The volcanic region here described, which has an area of about one hundred square miles, does not appear to have been at one and the same period of time the theatre of volcanic action over its whole extent. On the contrary, this action appears to have commenced at the southern border of the region in the Cantal, and travelling northwards, to have broken out in the Mont Dore region; finally terminating its outward manifestations among the craters and domes of the Puy de De. In a similar manner the volcanic eruptions of the Haute Loire and Ardhe, lying to the eastward, and separated from those of the Cantal by the granitoid ridge of the Montagnes de Margeride, belong to two successive periods referable very closely to those of the Mont Dore and the Puy de De groups.[5] The evidence in support of this view is very clear and conclusive; for, while the volcanic craters formed of ash, lapilli, and scori? together with the rounded domes of trachytic rock of which the Puy de De group is composed, preserve the form and surface indications of recently extinguished volcanoes, those which we may assume to have been piled up in the region of

Mont Dore and Cantal have been entirely swept away by prolonged rain and river action, and the sites of the ancient craters and cones of eruption are only to be determined by tracing the great sheets of lava up the sides of the valleys to their sources, generally situated at the culminating points of their respective groups. Other points of evidence of the great antiquity of the latter groups might be adduced from the extent of the erosion which has taken place in the sheets of lava having their sources in the vents of the Plomb du Cantal and of Mont Dore, owing to which, magnificent valleys, many miles in length and hundreds of feet in depth, have been cut out of these sheets of lava and their supporting rocks, whether granitic or lacustrine, and the materials carried away by the streams which flow along their beds. These points will be better understood when I come to give an account of the several groups; and in doing so I will commence with that of the Cantal.[6]

(d.) The Volcanoes of the Cantal.--The original crater-cones of this group have entirely disappeared throughout the long ages which have elapsed since the lava-streams issued forth from their internal reservoirs. The general figure of this group of volcanic mountains is that of a depressed cone, whose sides slope away in all directions from the central heights, which are deeply eroded by streams rising near the apex and flowing downwards in all directions towards the circumference of the mountain, where they enter the Lot, the Dordogne, and the Allier. The orifice of eruption was situated at the Plomb du Cantal, formed of solid masses of trachyte, which, owing, as Mr. Scrope supposes, to a high degree of fluidity, were able to extend to great distances in extensive sheets, and were afterwards covered by repeated and widely-spread flows of basalt; so that the trachyte towards the margin of the volcanic area becomes less conspicuous than the basalt by which it is more or less concealed from view, or overlapped. Extensive beds of tuff and breccia accompany the trachytic masses.

Magnificent sections of the rocks are laid open to view along the sides of the valleys, which are steep and rock-bound. Except towards the south-west, about Aurillac, where lacustrine strata overlie the granite, the platform from which rises the volcanic dome is composed of granitic or gneissose rocks. Accompanying the lava-streams are great beds of volcanic agglomerate, which Mr. Scrope considers to have been formed contemporaneously with the lava which they envelop, and to be due to torrents of water tumultuously descending the sides of the volcano at periods of eruption, and bearing down immense volumes of its fragmental ejecta in company with its lava-streams.[7] Nowhere throughout this region do beds of trachyte and basalt alternate with one another; in all cases the basalt is the newer of the two varieties of rock, and this is generally the case throughout the region here described.

(e.) Volcanoes of Mont Dore.--This mountain lies to the north of that of Cantal, and somewhat resembles it in general structure and configuration. Like Cantal, it is destitute of any distinct crater; all that is left of the central focus of eruption being the solidified matter which filled the throat of the original volcano, and which forms a rocky mass of lava, rising in its highest point, the Pic de Saucy, to an elevation (as given by Ramond) of 6258 feet above the level of the sea, thus exceeding that of the Plomb du Cantal by 128 feet. Its figure will be best understood by supposing seven or eight rocky summits grouped together within a circle of about a mile in diameter, from whence, as from the apex of an irregular and flattened cone, all the sides slope more or less rapidly downwards, until their inclination is gradually lost in the plain around. This dome-shaped mass has been deeply eroded on opposite sides by the valleys of the Dordogne and Chambon; while it is further furrowed by numerous minor streams.[8]

The great beds of volcanic rock, disposed as above stated, consist of prodigious layers of scori? pumice-stones, and detritus, alternating with beds of trachyte and basalt, which often descend in uninterrupted currents till they reach the granite platform, and then spread themselves for miles around. The sheets of basalt are found to stretch to greater distances than those of trachyte, and have flowed as far as 15 or 20 miles from their orifices of eruption; while in some cases, on the east and north sides, they have extended as far as 25 or 30 miles from the central height. On the other hand, a radius of about ten miles from the centre would probably include all the streams of trachyte;--so much greater has been the viscosity of the basalt over the latter rock. Some portions of these great sheets of lava, cut off by river valleys or eroded areas from the main mass of which they once formed a part, are found forming isolated terraces and plateaux either on the granitic platform, or resting on the fresh-water strata of the valley of the Allier, while in a northern direction they overspread a large portion of the granitic plateau from which rise the Puy de De and associated volcanic mountains. Still more remarkable are the cases in which these lava-streams have descended into the old river channels which drained the granitic plateau. Thus the current which took its origin in the Puy Gros descended into the valley of the Dordogne, while another stream invaded the gorge of Champeix on the eastern side.

The more ancient lava-streams just described are invaded by currents and surmounted by cones of eruption of more recent date, similar to those of the Puy de De group lying to the northward. Such cones and currents, amongst which are the Puy de Tartaret and that of Montenard, present exactly the same characters as those of this group, to which we shall return further on.

(f.) Volcanoes of the Haute Loire and Ardhe.--Separated by the valley of the Allier and the granitic ridge of La Margeride from the volcanic regions of Cantal and Mont Dore is another volcanic region of great extent, which reaches its highest elevation in the central points of Mont Mezen, attaining an elevation (according to Cordier) of 5820 feet, and formed of "clinkstone." The volcanic products of Mezen have been erupted from one central orifice of vast size, and consist mainly of extensive sheets of "clinkstone," a variety of trachytic lava, which have taken courses mainly towards the north-west and south-east. These great sheets, one of which appears to have covered a space more than 26 miles in length with an average breadth of 6 miles, thus overspreading an estimated area of 156 square miles, has been deeply eroded by streams draining into the Loire, along whose banks the rocks tower in lofty cliffs; while it has also suffered enormous denudation, by which outlying fragments are disconnected from the main mass, and form flat-topped hills and plateaux as far distant as Roche en Reigner and Beauzac, at the extreme distance (as stated above) of 26 miles from the source of eruption.

But even more remarkable than the above are the vast basaltic sheets which stretch away for a distance of 30 miles by Privas almost to the banks of the Rhe, opposite Montlimart. These have their origin amongst the clinkstone heights of Mont Mezen, and taking their course along the granitic plateau in a south-easterly direction, ultimately pass over on to the Jurassic and Cretaceous formations composing the plateau of the Coiron, which break off in vertical cliffs from 300 to 400 feet in height, surmounting the slopes that rise from the banks of the Ardhe and Escourtais rivers near Villeneuve de Bere. This is probably one of the most extensive sheets of basalt with which we are acquainted in the European area, and it is only a remnant of a vastly greater original sheet.[9]

(g.) Newer Volcanoes of the Haute Loire (the Velay and Vivarais).-- Subsequently to the formation of the lava-streams above described, and probably after the lapse of a lengthened period, the region of the Haute Loire and Ardhe became the scene of a fresh outburst of volcanic action, during which the surface of the older lavas, or of the fundamental granite, was covered by numerous crater-cones and lava-streams strewn along the banks of the Allier and of the Loire for many miles. These cones and craters are not quite so fresh as those of the Mont De group; those of the Haute Loire being slightly earlier in point of time, and, as Daubeny shows, belonging to a different system. So numerous are these more recent cones and craters that Scrope counted more than 150 of them, and probably omitted many.

The volcanic phenomena now described have a special interest as bearing on the question whether man was an inhabitant of this region at the time of these later eruptions. The question seems to be answered in the affirmative by the discovery of a human skull and several bones in the volcanic breccia of Mont Demise, in company with remains of the elephant (E. primigenius), rhinoceros (R. tichorhinus), stag, and other large mammifers. The discovery of these remains was made in the year 1844, and the circumstances were fully investigated and reported upon by M. Aymard, and afterwards by Mr. Poulett Scrope, upon whose mind no possible doubt of the fact remained. From what we now know of the occurrence of human remains and works of art in other parts of France and Europe, no surprise need be felt at the occurrence of human remains in company with some extinct mammalia in these volcanic tuffs, which belong to the Post-Pliocene or superficial alluvia antecedent to the historic period.[10]

(h.) Mont De Chain.--We now come to the consideration of the most recent of all the volcanic mountain groups of the region of Central France, that of the Puy de De, lying to the north of Mont Dore and Cantal. We have seen that there is almost conclusive evidence that man was a witness to the later volcanic outbursts of the Vivarais, and as these craters seem to be of somewhat earlier date than those of the Puy de De group, we cannot doubt that they were in active eruption when human beings inhabited the country, and not improbably within what is known as the Historic Period. No mention, however, is made either by Ceasar, Pliny, or other Roman writers of the existence of active volcanoes in this region. Ceasar, who was a close observer, and who carried the Roman arms into Auvergne, makes no mention of such; nor yet does the elder Pliny, who enumerated the known burning mountains of his day all over the Roman Empire. It is not till we come down to the fifth century of our era that we find any notices which might lead us to infer the existence of volcanic action in Central France. This is the well-known letter written by Sidonius Apollonarius, bishop of Auvergne, to Alcinus Avitus, bishop of Vienne, in which the former refers to certain terrific terrestrial manifestations which had occurred in the diocese of the latter. But, as Dr. Daubeny observes, this is no evidence of volcanic action in Auvergne, where Sidonius himself resided; the terrestrial disturbances above referred to may have been earthquake shocks of extreme severity.[11]

But although we have no reliably historical record of volcanic action amongst the mountains of the Mont De group, the fact that these are, comparatively, extremely recent will be evident to an observer visiting this district, and this conclusion is based on three principal grounds: first, because of the well-preserved forms of the original craters, though generally composed of very loose material, such as ashes, lapilli, and slag; secondly, because of the freshness of the lava-streams over whose rugged surfaces even a scanty herbage has in some places scarcely found a footing;[12] and thirdly, because the lava from the crater-cones has invaded channels previously occupied by the earlier lavas, or those which had been eroded since the overflow of the great basaltic sheets of Mont Dore. Still, as in the case of the valleys of Lake Aidot, of Channonat, and of Royat, these streams are sufficiently ancient to have given time for the existing rivers to have worn out in them channels of some depth, but bearing no comparison to the great valleys which had been eroded out of the more ancient lavas, such as those of the Coiron, of the Ardhe, and of the Dordogne and Chambon in the district of Mont Dore.

(i.) Dome-shaped Volcanic Hills.--I have previously (page 15) referred to the two classes of volcanic eminences to be found in the chain of the Puy de De; one indicated by the name itself, formed of a variety of trachytic lava called "domite," and of the form of a dome; the other, composed of fragmental matter piled up in the form of a crater or cup, often ruptured on one side by a stream of lava which has burst through the side, owing to its superior density. Of the former class the Puy de De and the Grand Sarcoui (see Fig. 18) are the most striking examples out of the five enumerated by Scrope, while there is a large number, altogether sixty-one, belonging to the latter class. These domes and crater-cones, as already stated, rise from a platform of granite, either directly or from one formed of the lava-sheets of the Mont Dore region, which in turn overlies the granitic platform. Of the nearly perfect craters there are the Petit Puy de De, lying partially against the northern flank of the greater eminence; the Puy de Cone, remarkable for the symmetry of its conical form, rising to a height of 900 feet from the plain; and the Puys de Chaumont and Thiolet lying to the north of the Puy de De. Of those to the south of this mount, two out of the three craters of the Puy de Barme and the Puy de Vichatel are perfect; but most of the crater-cones south of the Puy de De are breached. Some of the lava streams by which these craters were broken down flowed for long distances. That the lava followed the showers of ashes and lapilli forming the walls of the craters is rendered very evident in the case of the Puy de la Vache, whose lava-stream coalescing with those from the Puy de la Solas and Puy Noir, deluged the surrounding tracts and flowed down the Channonat Valley as far as La Roche Blanc in the Vale of Clermont. In the interior of the upper part of the crater still remaining may be seen the level (so to speak) to which the molten lava rose before it burst its barrier. This level is marked by a projecting platform of reddish or yellow material, rich in specular iron, apparently part of the frothy scum which formed on the surface of the lava and adhered to the side of the basin at the moment of its being emptied.

Space does not permit a fuller description of this remarkable assemblage of extinct volcanoes, and the reader must be referred for further details to the work of Mr. Scrope. I shall content myself with some further reference to the central figure in this grand chain, the Puy de De itself.

Ascent of the Puy de De.--On ascending by the winding path up the steep side of the mount, and on reaching the somewhat flattened summit, the first objects which strike the eye are the massive foundations of the Roman temple of Mercury; they are hewn out of solid grey lava, altogether different from the rock of the Puy de De itself, which must have been obtained from one of the lava-sheets of the Mont Dore group. To have carried these large blocks to their present resting-place must have cost no little labour and effort. The temple is supposed to have been surmounted by a colossal statue of the winged deity, visible from all parts of the surrounding country which was dedicated to his honour, and the foundations were only discovered a few years ago when excavating for the foundation of the observatory, which stands a little further on under the charge of Professor Janssen. On proceeding to the northern crest of the platform a wonderful view of the extinct craters and domes--about forty in number, and terminating in the Puy de Beauny, the most northerly member of the chain--is presented to the spectator. To the right is the Vale of Clermont and the rich valley of the Allier merging into the great plain of Central France. On the south side of the platform a no less remarkable spectacle meets the eye. The chain of Puys and broken craters stretches away southwards for a distance of nearly ten miles, while the horizon is bounded in that direction by the lofty masses of the Mont Dore, Cantal, and Le Puy ranges. Nor does it require much effort of the imagination to restore the character of the region when these now dormant volcanoes were in full activity, projecting showers of ashes and stones high into the air amidst flames of fire and vast clouds of incandescent gas and steam.

The material of which the Puy de De is formed consists of a light grey, nearly white, soft felsitic lava, containing crystals of mica, hornblende, and specular iron-ore. It is highly vesicular, and was probably extruded in a pasty condition from a throat piercing the granitic plateau and the overlying sheet of ancient lava of Mont Dore. It has been suggested that such highly felsitic and acid lavas as that of which the Puy de De, the Grand Sarcoui, and Cliersou are composed, may have had their origin in the granite itself, melted and rendered viscous by intense heat. Dr. E. Gordon Hull has suggested that the domite hills (owing to their low specific gravity) may have filled up pre-existing craters of ashes and scori?without rupturing them, as in the case of the heavier basaltic lavas, and then still continuing to be extruded, may have entirely enveloped them in its mass; so that each domite hill encloses within its interior a crater formed of ashes, stones, and scori? In the case of the Puy de De there is some evidence that the domite matter rests on a basis of ashes and scori? which may be seen in a few places around the base of the cone. It is difficult without some such theory as this to explain how a viscous mass was able to raise mountains some 2000 or 3000 feet above the surrounding plain.[13]

(j.) Sketch of the Volcanic History of Central France.--It now only remains to give a brief resum?of the volcanic history of this region as it may be gathered from the relations of the rocks and strata to the volcanic products, and of these latter to each other.

1st Stage.--It would appear that at the close of the Eocene period great terrestrial changes occurred. The bed of the sea was converted into dry land, the strata were flexured and denuded, and a depression was formed in the granitic floor of Central France, which, in the succeeding Miocene period, was converted into an extensive lake peopled by molluscs, fishes, reptiles, and pachyderms of the period.

2nd Stage.--Towards the close of the Miocene epoch volcanic eruptions commenced on a grand scale over the granitic platform in the districts now called Mont Dore, Cantal, and the Vivarais. Vast sheets of trachytic and basaltic lavas successively invaded the tracts surrounding the central orifices of eruption, now constituting the more ancient of the lava-sheets of the Auvergne region, and, invading the waters of the neighbouring lake, overspread the lacustrine deposits which were being accumulated therein. These volcanic eruptions probably continued throughout the Pliocene period, interrupted by occasional intervals of inactivity, and ultimately altogether ceased.

3rd Stage.--Towards the close of the Pliocene period terrestrial movements took place, owing to which the waters of the lake began to fall away, and the sheets of lava were subjected to great denudation. This process, probably accelerated by excessive rainfall during the succeeding Post-Pliocene and Pluvial periods, was continued until plains and extensive river-valleys were eroded out of the sheets of lava and their supporting granitic rocks and the adjoining lacustrine strata.

4th Stage.--A new outburst of volcanic forces marks this stage, during which the chain of the Puy de Dôme was thrown up on the west, and that of the newer cones of the Vivarais on the south-east of the lacustrine tract. The waters of the lake were now completely drained away through the channel of the Allier, and denudation, extending down to the present day, began over the area now forming the Vale of Clermont and adjoining districts. The volcanic action ultimately spent its force; and somewhere about the time of the appearance of man, the mammoth, rhinoceros, stag, and reindeer on the scene, eruptions entirely ceased, and gradually the region assumed those conditions of repose by which it is now physically characterised.

[1] The literature referring to this region is very extensive. Guettard in 1775, afterwards Faujas, published descriptions of the rocks of the Vivarais and Velay; and Desmarest's geological map, published in 1779, is a work of great merit. The district was afterwards described by Daubeny, Lyell, Von Buch, and others; but by far the most complete work is that of Scrope, entitled Volcanoes of Central France, containing maps and numerous illustrations, published in 1826, and republished in a more extended form in 1858; to this I am largely indebted.

[2] A monument to Pascal, erected by the citizens, occupies the centre of the square in Clermont. It will be remembered that Pascal verified the conclusions arrived at by Torricelli regarding the pressure of the atmosphere, by carrying a Torricellian tube to the summit of the Puy de De, and recording how the mercury continually fell during the ascent, and rose as he descended. This experiment was made in 1645.

[3] In this visit to Auvergne in the summer of 1880, the author was accompanied by his son, Dr. E. Gordon Hull, and Sir Robert S. Ball. On reaching the station at the summit of the ridge it seemed as if the volcanic fires had again been lighted, for the whole sky was aglow with the rays of the western sun.

[4] On the other hand, certain beds of ash and other volcanic ejecta occur in the uppermost strata of lake deposits of Limagne, so that these may indicate the commencement of the period of eruption, as suggested further on.

[5] Only very closely; for Mr. Scrope considers that the crater-cones of the chain of the Haute Loire give evidence of a somewhat earlier epoch of activity than those of the Puy de De, as they have undergone a greater amount of subaerial erosion.

[6] The extent of this river erosion has been clearly brought out by Scrope, and is admirably illustrated by several of his panoramic views, such as that in Plate IX. of his work.

[7] Scrope, loc. cit., p. 147.

[8] Scrope, loc. cit., p. 144.

[9] Scrope gives a view of these remarkable basaltic cliffs in Plate XII. of his work, from which the above account is taken. At one spot near the village of Le Gua there is a break in the continuity of the sheet.

[10] See Scrope, loc. cit., p. 181; also Appendix, p. 228. While there is no prim?facie reason for questioning the origin of the Demise skull, yet from what Lyell states in his Antiquity of Man, p. 196, it will be seen that he found it impossible to identify its position, or to determine beyond question that its interment was due to natural causes. But assuming this to be the case, he shows how the individual to whom it belonged might have been enveloped in volcanic tuff or mud showered down during the final eruption of the volcano of Demise. MM. Hert and Lartet, on visiting the locality, also failed to find in situ any exact counterpart of the stone now in the museum of Le Puy.

[11] See Daubeny, Volcanoes, p. 31.

[12] That is to say, the surfaces of the lava-streams are not at all, or only slightly, decomposed into soil suitable for the growth of plants, except in rare instances.

[13] E. G. Hull, "On the Domite Mountains of Central France," Scien. Proc. Roy. Dublin Society, July 1881, p. 145. Dr. Hull determined the density of the domite of the Puy de De to be 2.5, while that of lava is about 3.0.

CHAPTER VII.

THE VOLCANIC DISTRICT OF THE RHINE VALLEY.

The region bordering the Rhine along both its banks above Bonn, and extending thence along the valley of the Moselle and into the Eifel, has been the theatre of active volcanic phenomena down into recent times, but at the present day the volcanoes are dormant or extinct.

(a.) Geological Structure.--The fundamental rocks of this region belong to the Silurian, Devonian, and Carboniferous systems, consisting of schists, grits, and limestones, with occasional horizontal beds of Miocene sandstone and shale with lignite, resting on the upturned edges of the older rocks. Scattered over the greater part of the district here referred to are a number of conical eminences, often with craters, the bottoms of which are usually sunk much below the present level of the country, and thus receiving the surface drainage, have been converted into little lakes called "maars," differing from ordinary lakes by their circular form and the absence of any apparent outlet for their waters.[1]

But before entering into details, it may be desirable to present the reader with a short outline of the physical history of the region (which has been ably done by Dr. Hibbert in his treatise, to which I have already referred), so as to enable him better to understand the succession of physical events in its volcanic history.

(b.) Physical History.--From the wide distribution of stratified deposits of sand and clay at high levels on both banks of the Rhine north of the Moselle, it would appear that an extensive fresh-water basin, which Dr. Hibbert calls "The Basin of Neuwied," occupied a considerable tract on both banks, in the centre of which the present city of Neuwied stands. This basin was bounded towards the south by the slopes of the Hedsruck and Taunus, which at the time here referred to formed a continuous chain of mountains. (Fig. 20.) To the south of this chain lay the Tertiary basin of Mayence, which was connected at an early period--that of the Miocene--with the waters of the ocean, as shown by the fact that the lower strata contain marine shells; these afterwards gave place to fresh-water conditions. The basin of Neuwied was bounded towards the north by a ridge of Devonian strata which extended across the present gorge of the Rhine between Andernach and Linz, and to the north of this barrier lay another more extensive fresh-water basin, that of Cologne. From this it will be seen that the Rhine, as we now find it, had then only an infantile existence; in fact, its waters to the south of the Hedsruck ridge drained away towards the south. But towards the commencement of the Pliocene period the barriers of the Hedsruck and Taunus, as also that of the Linz, were broken through, and the course of the waters was changed; and thus gradually, as the river deepened its bed, the waters were drained off from the great lakes.[2] This rupture of the barriers may have been due, in the first instance, to the terrestrial disturbances accompanying the volcanic eruptions of the Eifel and Siebengebirge, though the erosion of the gorges at

Bingen and at Linz to their present depth and dimensions is of course due to prolonged river action. It was about the epoch we have now arrived at--viz., the close of the Miocene--that volcanic action burst forth in the region of the Lower Rhine. It is probable that this action commenced in the district of the Siebengebirge, and afterwards extended into that of the Moselle and the Eifel, the volcanoes of which bear evidence of recent date. Layers of trachytic tuff are interstratified with the deposits of sand, clay, and lignite of the formation known as that of the Brown Coal--of Miocene age--which underlies nearly the whole of the volcanic district on both sides of the Rhine near Bonn,[3] thus showing that volcanic action had already commenced in that part to some extent; but it does not appear from Dr. Hibbert's statement that any such fragments of eruptive rock are to be found in the strata which were deposited over the floor of the Neuwied basin.[4] It will be recollected that the epoch assigned for the earliest volcanic eruptions of Auvergne was that here inferred for those of the Lower Rhine--viz., the close of the Miocene stage--and from evidence subsequently to be adduced from other European districts, it will be found that there was a very widely spread outburst of volcanic action at this epoch.

(c.) The Range of the Siebengebirge.--This range of hills--formed of the older volcanic rocks of the Lower Rhine--rises along the right bank of this noble river opposite Bonn, where it leaves the narrow gorge which it traverses all the way from Bingen, and opens out on the broad plain of Northern Germany. The range consists of a succession of conical hills sometimes flat-topped--as in the case of Petersberg; and at the Drachenfels, near the centre of the range it presents to the river a bold front of precipitous cliffs of trachyte porphyry. The sketch (Fig. 21) here presented was taken by the author in 1857 from the old extinct volcano of Roderberg, and will convey, perhaps, a better idea of the character of this picturesque range than a description. The Siebengebirge, although appearing as an isolated group of hills, is in reality an offshoot from the range of the Westerwald, which is connected with another volcanic district of Central Germany known as the Vogelsgebirge. The highest point in the range is attained in the Lohrberg, which rises 1355 feet above the sea; the next, the Great Trakeberg, 1330 feet; and the next, Great Oelberg, 1296 feet.

The range consists mainly of trachytic rocks--namely, trachyte-conglomerate, and solid trachyte, of which H. von Dechen makes two varieties--that of the

Drachenfels, and that of the Wolkenburg. But associated with these highly-silicated varieties of lava--and generally, if not always, of later date--are basaltic rocks which cap the hills of Petersberg, Nonnenstrom, Gr. and Ll. Oelberg, Gr. Weilberg, and Ober Dollendorfer Hardt. The question whether there is a transition from the one variety of volcanic rock into the other, or whether each belongs to a distinct and separate epoch of eruption, does not seem to be very clearly determined. Mr. Leonard Horner states that it would be easy to form a suite of specimens showing a gradation from a white trachyte to a black basalt;[5] but we must recollect that when Mr. Horner wrote, the microscopic examination of rocks by means of thin sections was not known or practised, and an examination by this process might have proved that this apparent transition is unreal. According to H. von Dechen, there are sheets of basalt older than the greater mass of the brown coal formation, and others newer than the trachyte;[6] while dykes of basalt traversing the trachytic lavas are not uncommon.[7]

The trachyte-conglomerate--which seems to be associated with the upper beds of the brown coal strata--is traversed by dykes of trachyte of later date; and though it is difficult to trace the line between the two varieties of this rock on the ground, Dr. von Rath has recognised the general distinction between them, which consists in the greater abundance of hornblende and mica in the trachyte of the Wolkenburg than in that of the Drachenfels.

The trachyte of the Drachenfels was probably the neck of a volcano which burst through the fundamental schists of the Devonian period. It is remarkable for the large crystals of sanidine (glassy felspar) which it contains, and has a rude columnar structure.

The absence of any clearly-defined craters of eruption, such as are to be found in the Eifel district and on the left bank of the Rhine--as, for example, in the case of the Roderberg--may be regarded as sufficient evidence that this range is of comparatively high antiquity. It seems to bear the same relation to the more modern craters of the Eifel and Moselle that the Mont Dore and Cantal volcanoes do to those of the Puy de De. In both cases, denudation carried on throughout perhaps the Pliocene and Post-Pliocene periods down to the present day has had the effect of demolishing the original craters; so that what we now observe as forming these ranges are the consolidated columns of original molten matter which filled the throats of the old volcanoes, or the sheets of lava which were extruded from them, but are now probably much reduced in size and extent.

Having thus given a description of the older volcanic range on the right bank of the Rhine, we shall cross the river in search of some details regarding the more recent group of Rhenish volcanoes, commencing with that of the Roderberg, a remarkable hill a few miles south of Bonn, from which the view of the Seven Mountains was taken.

(d.) The Roderberg.--This crater, which was visited by the author in 1857, is about one-fourth of a mile in diameter, and is in the form of a cup with gentle slopes on all sides. In its centre is a farmhouse surrounded by corn-fields. The general section through the hill is represented above (Fig. 22).

The flanks on the north side are composed of loose quartzose gravel (gerolle), a remnant of the deposits formed around the margin of the "Basin of Neuwied" described above (p. 114). This gravel is found covering the terraces of the brown coal formation several hundred feet above the Rhine. Besides quartz-pebbles, the deposit contains others of slate, grit, and volcanic rock. On reaching the edge of the crater we find the gravel covered over by black and purple scoria or slag the superposition of the scoria on the gravel being visible in several places, showing that the former is of more recent origin. On the opposite side of the crater, overlooking the Rhine, we find the cliff of Rolandsec composed of hard vesicular lava, rudely prismatic, and extending from the summit of the hill to its base, about 250 feet below. This is the most northerly of the group of the Eifel volcanoes.

(e.) District of the Rivers Brul and Nette.--The volcanic region of the Lower Eifel, drained by these two principal streams which flow into the Rhine, will amply repay exploration by the student of volcanic phenomena, owing to the variety of forms and conditions under which these present themselves within a small space. The fundamental rock is slate or grit of Devonian age, furrowed by numerous valleys, often richly wooded, and diversified by conical hills of trachyte; or by crater-cones, formed of basalt or ashes, sometimes ruptured on one side, and occasionally sending forth streams of lava, as in the cases of the Perlinkopf, the Bausenberg, and the Engelerkopf. The district attains its greatest altitude in the High Acht (Der Hohe Acht), an isolated cone of slate capped by basalt with olivine, and reaching a level of 2434 Rhenish feet.[8]

(f.) The Laacher See.--It would be impossible in a work of this kind to attempt a detailed description of the Eifel volcanoes, often of a very complex character and obscure physical history, as in the case of the basin of Rieden, where tufaceous deposits, trachytic and basaltic lavas and crater-cones, are confusedly intermingled, so that I shall confine my remarks to the deservedly famous district of the Laacher See, which I had an opportunity of personally visiting some years since.[9]

The Laacher See is a lake of an oval form, over an English mile in the shorter diameter, and surrounded by high banks of volcanic sand, gravel, and scori? except on the east side, where cliffs of clay-slate, in a nearly vertical position, and striking nearly E.W., may be observed. Its depth from the surface of the water is 214 feet.[10] The ashes of the encircling banks contain blocks of slate and lava which have been torn from the sides of the orifice or neck of the volcano and blown into the air; and there can be no doubt that the ashes and volcanic gravel is the result of very recent eruptions.

At the east side of the lake we find a stream of scoriaceous lava of a purple or reddish colour, highly vesicular, and containing crystals of mica; but the most important lava-stream is that which has taken a southerly direction from the crater of the Laacher See towards Nieder Mendig and Mayen, for a distance of about six miles. This great stream is covered throughout half its distance by beds of volcanic ash and lapilli, but emerges into the air at a distance of about two miles from the edge of the crater (see Fig. 23), and was formerly extensively quarried in underground caverns for millstones. Here the rock is a vesicular trachyte, of a greyish colour, solidified in vertical columns of hexagonal form, about four feet in diameter, and traversed by transverse joint planes. These quarries have been worked from the time of the Roman occupation of the country; and, before the introduction of iron or steel rollers for grinding corn, millstones were exported to all parts of Europe and the British Isles from this quarry.[11]

The district around the Laacher See is covered by laminated ejecta of the old volcano, probably of subaerial origin, through which bosses of the fundamental slate peer up at intervals, while the surface is diversified by several truncated cones.

(g.) Trass of the Brul Valley.--The Brul Valley, which unites with that of the Rhine at the town of that name, and drains the northern side of the volcanic region, has always been regarded with much interest by travellers for the presence of a deposit of "trass" with which it is partially filled. The origin of this valley was pre-volcanic, as it is hewn out of the slaty rocks of the district. But at a later period it became filled with volcanic mud (tuffstein), out of which the stream has made for itself a fresh channel. The source of this mud is considered by Hibbert[12] to have been the old volcano of the Lummerfeld, which, after becoming dormant, was filled with water, and thus became a lake. At a subsequent period, however, a fresh eruption took place near the edge of the lake, resulting in the remarkable ruptured crater known as the Kunkskife, which rises about four miles to the north of the Laacher See. The eruptions of this volcano appear to have displaced the mud of the Lummerfeld, causing it to flow down into the deep gorge which it completely filled, as stated above.

On walking down the valley one may sometimes see the junction of the tuff with the slate-rock which enfolds it. The tuff consists of white felspathic mud, with fragments of slate and lava, reaching a depth in some places of 150 feet. After it has been quarried it is ground in mills, and used for cement stone under the name of trass. It is said to resemble the volcanic mud by which Herculaneum was overwhelmed during the first eruption of Vesuvius, and which was produced by the torrents of rain mixing with the ashes as they were blown out of the volcano.

Sufficient has probably now been written regarding the dormant, or recently extinct, volcanic districts of Europe to give the reader a clear idea regarding their nature and physical structure. Other districts might be added, such as those of Central Germany, Hungary, Transylvania, and Styria; but to do so would be to exceed the proposed limits of this work; and we may therefore pass on to the consideration of the volcanic region of Syria and Palestine, which adjoins the Mediterranean district we have considered in a former page.

[1] Daubeny, loc. cit., p. 71. The geology of this region has had many investigators, of whom the chief are Steininger, Erloschenen Vulkane in der Eifel (1820); Hibbert, Extinct Volcanoes of the Basin of Neuwied, 1832; gerath, Das Gebirge im Rheinland, etc., 4 vols.; Horner, "On the Geology of Bonn,"

Transactions of the Geological Society, London, vol. iv.

[2] The views of Dr. Hibbert are not inconsistent with those of the late Sir A. Ramsay, on "The Physical History of the Valley of the Rhine," Quart. Jour. Geol. Soc., vol. xxx. (1874).

[3] Von Dechen, Geog. Beschreib. des Siebengebirges am Rhein (Bonn, 1852).

[4] Hibbert, loc. cit., p. 18.

[5] Horner, "Geology of Environs of Bonn," Transactions of the Geological Society, vol. iv., new series.

[6] H. von Dechen, Geog. Fehrer in das Siebengebirge am Rhein (Bonn, 1861).

[7] Ibid., p. 191.

[8] Dr. Hibbert's work is illustrated by very carefully drawn and accurate views of some of the old cones and craters of this district, accompanied by detailed descriptions.

[9] The lava of Schorenberg, near Rieden, is interesting from the fact, stated by Zirkel, that it contains leucite, nosean, and nephelin.--Die Mikros. Beschaf. d. Miner. u. Gesteine, p. 154 (1873).

[10] Hibbert, loc. cit., p. 23.

[11] At the time of the author's visit the underground caverns, which are deliciously cool in summer, were used for the storage of the celebrated beer brewed by the Moravians of Neuwied.

[12] Hibbert, loc. cit., p. 129.

PART III.

DORMANT OR MORIBUND VOLCANOES OF OTHER PARTS OF THE WORLD.

CHAPTER I.

DORMANT VOLCANOES OF PALESTINE AND ARABIA.

(a.) Region east of the Jordan and Dead Sea.--The remarkable line of country lying along the valley of the Jordan, and extending into the great Arabian Desert, has been the seat of extensive volcanic action in prehistoric times. The specially volcanic region seems to be bounded by the depression of the Jordan, the Dead Sea, and the Arabah as far south as the Gulf of Akabah; for, although Safed, lying at the head of the Sea of Galilee on the west of the Jordan valley, is built on a basaltic sheet, and is in proximity to an extinct crater, its position is exceptional to the general arrangement of the volcanic products which may be traced at intervals from the base of Hermon into Central Arabia, a distance of about 1000 miles.[1]

The tract referred to has been described at intervals by several authors, of whom G. Schumacher,[2] L. Lartet,[3] Canon Tristram,[4] M. Niebuhr,[5] and C. M. Doughty[6] may be specially mentioned in this connection.

The most extensive manifestations of volcanic energy throughout this long tract of country appear to be concentrated at its extreme limits. At the northern extremity the generally wild and rugged tract of the Jaul 鈔 and Haur 鈔, called in the Bible Trachonitis, and still farther to the eastward the plateau of the Lejah, with its row of volcanic peaks sloping down to the vast level of Bashan, is covered throughout nearly its whole extent by great sheets of basaltic lava, above which rise at intervals, and in very perfect form, the old crater-cones of eruption. A similar group of extinct craters with lava-flows has been described and figured by a recent traveller, Mr. C. M. Doughty, in parts of Central Arabia. The general resemblance of these Arabian volcanoes to those of the Jaul 鈔 is unquestionable; and as they are connected with each other by sheets of basaltic lava at intervals throughout the land of Moab, it is tolerably certain that the volcanoes lying at either end of the chain belong to one system, and were contemporaneously in a state of activity.

(b.) Geological Conditions.--Before entering any further into particulars regarding the volcanic phenomena of this region, it may be desirable to give a short account of its geological structure, and the physical conditions amongst which the igneous eruptions were developed.

Down to the close of the Eocene period the whole region now under consideration was occupied by the waters of the ocean. The mountains of Sinai were islands in this ocean, which had a very wide range over parts of Asia, Africa, and Europe. But at the commencement of the succeeding Miocene stage the crust was subjected to lateral contraction, owing to which the ocean bed was upraised. The strata were flexured, folded, and often faulted and fissured along lines ranging north and south, the great fault of the Jordan-Arabah valley being the most important. At this period the mountains of the Lebanon, the table-lands of Judea and of Arabia, formed of limestone, previously constituting the bed of the ocean during the Eocene and Cretaceous periods, were converted into land surfaces. Along with this upheaval of the sea-bed there was extensive denudation and erosion of the strata, so that valleys were eroded over the subaerial tracts, and the Jordan-Arabah valley received its primary form and outline.

Up to this time there does not appear to have been any outbreak of volcanic forces; but with the succeeding Pliocene period these came into play, and eruptions of basaltic lava took place along rents and fissures in the strata, while craters and cones of slag, scori and ashes were thrown up over the region lying to the east of the Sea of Galilee and the sources of the Jordan on the one hand, and the central parts of the great Arabian Desert on the other. These eruptions, probably intermittent, continued into the succeeding Glacial or Pluvial period, and only died out about the time that the earliest inhabitants appeared on the scene.

(c.) The Jaul and Haur.--This tract is bounded by the valley of the Jordan and the Sea of Galilee on the west, from which it rises by steep and rocky declivities into an elevated table-land, drained by the Hieromax, the Nahr er Rukk, and other streams, which flow westwards into the Jordan along deep channels in which the basaltic sheets and underlying limestone strata are well laid open to view.

On consideration it seems improbable that the great sheets of augitic lava, such as cover the surface of the land of Bashan, are altogether the product of the volcanic mountains which appear to be confined to special districts in this wide area. Some of the craters do indeed send forth visible lava-streams, but they are insignificant as compared with the general mass of the plateau-basalts; and the crater-cones themselves appear in some cases to be posterior to the platforms of basalt from which they rise. It is very probable, therefore, that the lavas of this region have, in the main, been extruded from fissures of eruption at an early period, and spread over the surface of the country in the same manner as those of the Snake River region, and the borders of the Pacific Ocean of North America, and possibly of the Antrim Plateau in Ireland, afterwards to be described.

The volcanic hills which rise above the plateau are described in detail by Schumacher. Of these, Tell Ab?Ned is the largest in the Jaul. It reaches an elevation of 4132 feet above the Mediterranean Sea, and 1710 feet above the plain from which it rises; the circumference of its base is three miles, and the rim of the crater itself, which is oval in form, is 1331 yards in its larger diameter. The interior is cultivated by Circassians, and is very fruitful; the walls descend at an angle of about 30?on the inside, the exterior slope of the mountain being about 22? The cone seems to be formed chiefly of scori? and the lava-stream, which issues forth from the interior, forms a frightfully stony and lacerated district.[7]

Another remarkable volcano is the Tell Ab?en Ned?(Fig. 24). This is a double crater, with a cone (probably of cinders) rising from the interior of one of them. The highest point of the rim of one of the craters reaches a level of 4042 feet above the sea. A lava-stream issues forth from Ab?en Ned? and unites with another from a neighbouring volcano.

Tell el Ahm is a ruptured crater of imposing aspect, reaching an elevation of 4060 feet, and sending forth a lava-current, which falls in regular terraces from the outlet towards the west and north.

The ruptured crater of Tell el Akkasheh, which reaches a height of 3400 feet, has a less forbidding aspect than the greater number of the extinct volcanoes of this region, owing to the fact that its sides are covered by oaks, which attain to magnificent proportions along the summit. Numerous other volcanic hills occur in this district, but the most remarkable is that called Tell el Farras (the Hill of the Horse). It is an isolated mountain, visible from afar, and reaches an elevation of 3110 feet, or nearly 800 feet above the surrounding plain. The oval crater of this volcano opens towards the north, and has a depth of 108 feet below the edge, with moderately steep sloping sides (17?32?, while the slope of the exterior, at first steep, gradually lessens to

20?21? These slopes are covered with reddish or yellowish slag. The above examples will probably suffice to afford the reader a general idea of the size and form of the volcanoes in this little known region.

It has been stated above that the great lava-floods have probably been poured forth intermittently. The statement receives confirmation from the observations of Canon Tristram, made in the valley of the Yarm.[8] This impetuous torrent rushes down a gorge, sometimes having limestone on one side and a wall of basalt on the other. This is due to the fact that the river channel had been eroded before the volcanic eruptions had commenced; but on the lava-stream reaching the channel, it naturally descended towards the valley of the Jordan along its bed, displacing the river, or converting it into clouds of steam. Subsequently the river again hewed out its channel, sometimes in the lava, sometimes between this rock and the chalky limestone. But, in addition to this, it has been observed that there is a bed of river gravel interposed between two sheets of basalt in the Yarm ravine; showing that after the first flow of that molten rock the river reoccupied its channel, which was afterwards invaded by another molten lava-stream, into which the waters have again furrowed the channel which they now occupy. The basaltic sheets descend under the waters of the Sea of Galilee on the east side, and were probably connected with those of Safed, crossing the Jordan valley north of that lake; owing to this the waters of the Lake of Merom (Huleh) were pent up, and formerly covered an extensive tract, now formed of alluvial deposits.

(d.) Land of Moab.--Proceeding southwards into the Land of Moab, the volcanic phenomena are here of great interest. Extensive sheets of basaltic lava, described as far back as 1807 by Seetzen, and more recently by Lartet and Tristram, are found at intervals between the W 鈷 ies Mojib (Arnon) and Haidan. On either side of the Mojib, cliffs of columnar basalt are seen capping the beds of white Cretaceous limestone, while a large mass has descended into the W. Haidan between cliffs of limestone and marl on either hand.

Around Jebel Attarus--a dome-shaped hill of limestone--a sheet of basaltic lava has been poured, and has descended the deep gorge of the Zerka, which enters the Dead Sea some 2000 feet below. This gorge had been eroded before the basaltic eruption, so that the stream of molten lava took its course down the bed of this stream to the water's edge, and grand sections have been laid bare by subsequent erosion along the banks. Pentagonal columns of black basalt form perpendicular walls, first on one side, then on the other; while considerable masses of scori? peperino, and breccia appear at the head of the glen, probably marking the orifice of eruption. Other eruptions of basalt occur, one at Mountar ez Zara, to the south of Zerka, and another at Wady Ghuweir, near the north-eastern end of the Dead Sea. There are no lava-streams on the western side of the Ghor, or of the Dead Sea.[9]

The outburst of the celebrated thermal springs of Callirrho? together with nine or ten others, along the channel of the Zerka, is a circumstance which cannot be dissociated from the occurrence of basaltic lava at this spot. In a reach of three miles, according to Tristram, there are ten principal springs, of which the fifth in descent is the largest; but the seventh and eighth, about half a mile lower down, are the most remarkable, giving forth large supplies of sulphurous water. The tenth and last is the hottest of all, indicating a temperature of 143?Fahr. Thus it would appear that the heat increases with the depth from the upper surface of the table-land; a result which might be expected, supposing the heated volcanic rocks to be themselves the source of the high temperature. To a similar cause may be attributed the hot-springs of Hammath, near Tiberias, and those of the Yarm near its confluence with the Jordan. Some of these and other springs break out along, or near, the line of the great Jordan-Arabah fault which ranges throughout the whole extent of this depression, from the base of Hermon to the Gulf of Akabah, generally keeping close to the eastern margin of the valley.

(e.) The Arabian Desert.--The basaltic lava-floods occupy a very large extent of the Arabian Desert, from El Hisma (lat. 27?35' N.) to the neighbourhood of Mecca on the south, a distance of about 440 miles, with occasional intervals. The lava-sheets are called "Harras" (or "Harrat"), one of which, Harrat Sfeina, terminates about ten miles north of Mecca. The lava-sheets rest sometimes on the red sandstone, at other times, on the granite and other crystalline rocks of great geological antiquity. In addition to the sheets of basalt, numerous crater-cones rise from the basaltic platform at a level of 5000 feet above the sea, and two volcanic mountains, rising far to the west of the principal range, called respectively Harr Jeheyma and H. Rodwa, almost overlook the coast of the Red Sea.[10]

(f.) Age of the Volcanic Eruptions.--It is very clear that the first eruptions, producing the great basaltic sheets of Moab and Arabia, occurred after the principal features of the country had been developed. The depression of the Jordan-Arabah valley, the elevation of the eastern side of this valley along the great fault line, and the channels of the principal tributary streams, such as those of the Yarm and Zerka, all these had been eroded out before they were invaded by the molten streams of lava. Now, as these physical features were developed and sculptured out during the Miocene period, as I have elsewhere shown to be the case,[11] we may with great probability refer the volcanic eruptions to the geological epoch following--namely, the Pliocene. How far downwards towards the historic period the eruptions continued is not so certain. Dr. Daubeny, quoting several passages from the Old Testament prophets,[12] says it might be inferred that volcanoes were in activity even so late as to admit of their being included within the limits of authentic history. The poetic language and imagery used in these passages by the prophets certainly lends a probability to this view, but nothing more. On the other hand, these regions have suffered through many centuries from the secondary effects of seismic action and subterranean forces, and earthquake shocks have laid in ruins the great temples and palaces of Palmyra, Baalbec, and other cities of antiquity. The same uncertainty regarding the time at which volcanic action died out, with reference to the appearance of man on the scene, hangs over the region of Arabia and Syria, as we have seen to be the case in reference to the extinct volcanoes of Auvergne, the Eifel, and the Lower Rhine. In all these cases the commencement and close of eruptive action appear to have been very much about the same period--namely, the Miocene period on the one hand, and that at which man entered upon the scene on the other; but in the case of Syria and Western Palestine, the close of the volcanic period may have been somewhat more than 2000 B.C.

[1] Lake Phiala, near the Lake of Huleh, is also situated to the west of the Jordan valley. Its origin, according to Tristram, is volcanic.

[2] Schumacher, "The Jaul," Quarterly Statement of the Palestine Exploration Fund, 1886 and 1888; and Across the Jordan, London, 1886.

[3] Lartet, Voyage d'Exploration de la mer Morte (G 閘 logie), Paris, 1880.

[4] Tristram, Land of Moab, London, 1873; and Land of Israel, 1866.

[5] Niebuhr, Beschreibung von Arabien, 1773.

[6] C. M. Doughty, Arabia Deserta, 2 vols., 1888. A generalised account of this volcanic region by the author will be found in the "Memoir on the Physical Geology of Arabia Petra, and Palestine," Palestine Exploration Fund, 1887.

[7] Schumacher, loc. cit., p. 248.

[8] Land of Israel, p. 461.

[9] "Geology of Arabia Petra, and Palestine," Memoirs of the Palestine Exploration Fund, p. 95.

[10] Doughty, loc. cit., vol. i., plate vi., p. 416. An excellent geological sketch map accompanies this work.

[11] "Memoir of the Geology of Arabia Petra, and Palestine," chap. vi. p. 67.

[12] Nahum, i. 5, 6; Micah, i. 3, 4; Isaiah, lxiv. 1-3; Jeremiah, l. 25.

CHAPTER II.

THE VOLCANIC REGIONS OF NORTH AMERICA.

(a.) Contrast between the Eastern and Western Regions.--In no point is there a more remarkable contrast between the physical structure of Eastern and Western America than in the absence of volcanic phenomena in the former and their prodigious development in the latter. The great valley of the Mississippi and its tributaries forms the dividing territory between the volcanic and non-volcanic areas; so that on crossing the high ridges in which the western tributaries of America's greatest river have their sources, and to which the name of the "Rocky Mountains" more properly belongs, we find ourselves in a region which, throughout the later Tertiary times down almost to the present day, has been the scene of volcanic operations on the grandest scale; where lava-floods have been poured over the country through thousands of square miles, and where volcanic cones, vying in magnitude with those of Etna, Vesuvius, or Hecla, have established themselves. This region, generally known as "The Great Basin," is bounded on the west by the "Pacific Range" of mountains, and includes portions of New Mexico, Arizona, California, Nevada, Utah, Colorado, Idaho, Oregon, Wyoming, Montana, and Washington. To the south it passes into the mountainous region of Mexico, also highly volcanic; and thence into the ridge of Panama and the Andes. It cannot be questioned but that the volcanic nature of the Great Basin is due to the same causes which have originated the volcanic outbursts of the Andes; but, from whatever cause, the volcanic forces have here entered upon their secondary or moribund stage. In the Yellowstone Valley, geysers, hot springs, and fumaroles give evidence of this condition. In other districts the lava-streams are so fresh and unweathered as to suggest that they had been erupted only a few hundred years ago; but no active vent or crater is to be found over the whole of this wide region. A few special districts only can here be selected by way of illustration of its special features in connection with its volcanic history.

(b.) The Plateau Country of Utah and Arizona.--This tract, which is drained by the Colorado River and its tributaries, is bounded on the north by the Wahsatch range, and extends eastwards to the base of the Sierra Nevada. Round its margin extensive volcanic tracts are to be found, with numerous peaks and truncated cones--the ancient craters of eruption--of which Mount San Francisco is the culminating eminence. South of the Wahsatch, and occupying the high plateaux of Utah, enormous masses of volcanic products have been spread over an area of 9000 square miles, attaining a thickness of between 3000 and 4000 feet. The earlier of these great lava-floods appear to have been trachytic, but the later basaltic; and in the opinion of Captain Dutton, who has described them, they range in point of time from the Middle Tertiary (Miocene) down to comparatively recent times.

(c.) The Grand Canyon.--To the south of the high plateaux of Utah are many minor volcanic mountains, now extinct; and as we descend towards the Grand Canyon of Colorado we find numerous cinder-cones scattered about at intervals near the cliffs.[1] Extensive lava-fields, surmounted by cinder-cones, occupy the plateau on the western side of the Grand Canyon; and, according to Dutton, the great sheets of basaltic lava, of very recent age, which occupy many hundred square miles of desert, have had their sources in these cones of eruption.[2] Crossing to the east of the Grand Canyon, we find other lava-floods poured over the country at intervals, surmounted by San Francisco--a volcanic mountain of the first magnitude--which reaches an elevation, according to Wheeler, of 12,562 feet above the ocean. It has long been extinct, and its summit and flanks are covered with snow-fields and glaciers. Other parts of Arizona are overspread by sheets of basaltic lava, through which old "necks" of eruption, formed of more solid lava than the sheets, rise occasionally above the surface, and are prominent features in the landscape.

Further to the eastward in New Mexico, and near the margin of the volcanic region, is another volcanic mountain little less lofty than San Francisco, called Mount Taylor, which, according to Dutton, rises to an elevation of 11,390 feet above the ocean, and 8200 feet above the general level of the surrounding plateau of lava. This mountain forms the culminating point of a wide volcanic tract, over which are distributed numberless vents of eruption. Scores of such vents--generally cinder-cones--are visible in every part of the plateau, and always in a more or less dilapidated condition.[3] Mount Taylor is a volcano, with a central pipe terminating in a large crater, the wall of which was broken down on the east side in the later stage of its history.

[Illustration: Fig. 25.--Mount Shasta (14,511 feet), a snow-clad volcanic cone in California, with Mount Shastina, a secondary crater, on the right; the valley between is filled with glacier-ice.--(After Dutton).]

(d.) California.--Proceeding westwards into California, we are again confronted with volcanic phenomena on a stupendous scale. The coast range of mountains, which branches off from the Sierra Nevada at Mount Pinos, on the south, is terminated near the northern extremity of the State by a very lofty mountain of volcanic origin, called Mount Shasta, which attains an elevation of 14,511 feet (see Fig. 25). This mountain was first ascended by Clarence King in 1870,[4] and although forming, as it were, a portion of the Pacific Coast Range, it really rises from the plain in solitary grandeur, its summit covered by snow, and originating several fine glaciers.

The summit of Mount Shasta is a nearly perfect cone, but from its north-west side there juts out a large crater-cone just below the snow-line, between which and the main mass of the mountain there exists a deep depression filled with glacier ice. This secondary crater-cone has been named Mount Shastina, and round its inner side the stream of glacier ice winds itself, sometimes surmounting the rim of the crater, and shooting down masses of ice into the great caldron. The length of this glacier is about three miles, and its breadth about 4000 feet. Another very lofty volcanic mountain is Mount Rainier, in the Washington territory, consisting of three peaks of which the eastern possesses a crater very perfect throughout its entire circumference. This mountain appears to be formed mainly of trachytic matter. Proceeding further north into British territory, several volcanic mountains near the Pacific Coast are said to exhibit evidence of activity. Of these may be mentioned Mount Edgecombe, in lat. 57?3; Mount Fairweather, lat. 57?20 which rises to a height of 14,932 feet; and Mount St. Elias, lat. 60?5, just within the divisional line between British and Russian territory, and reaching an altitude of 16,860 feet. This, the loftiest of all the volcanoes of the North American continent, except those of Mexico, may be considered as the connecting link in the volcanic chain between the continent and the Aleutian Islands.[5]

(e.) Lake Bonneville.--Returning to Utah we are brought into contact with phenomena of special interest, owing to the inter-relations of volcanic and lacustrine conditions which once prevailed over large tracts of that territory. The present Great Salt Lake, and the smaller neighbouring lakes, those called Utah and Sevier, are but remnants of an originally far greater expanse of inland water, the boundaries of which have been traced out by Mr. C. K. Gilbert, and described under the name of Lake Bonneville.[6] The waters of this lake appear to have reached their highest level at the period of maximum cold of the Post-Pliocene period, when the glaciers descended to its margin, and large streams of glacier water were poured into it. Eruptions of basaltic lava from successive craters appear to have gone on before, during, and after the lacustrine epochs; and the drying up of the waters over the greater extent of their original area, now converted into the Sevier Desert, and their concentration into their present comparatively narrow basins, appears to have proceeded pari passu with the gradual extinction of the volcanic outbursts. Two successive epochs of eruption of basalt appear to have been clearly established--an earlier one of the "Provo Age," when the lava was extruded from the Tabernacle craters, and a later epoch, when the eruptions took place from the Ice Spring craters. The oldest volcanic rock appears to be rhyolite, which peers up in two small hills almost smothered beneath the lake deposits. Its eruption was long anterior to the lake period. On the other hand, the cessation of the eruptions of the later basaltic sheets is evidently an event of such recent date that Mr. Gilbert is led to look forward to their resumption at some future, but not distant, epoch. As he truly observes, we are not to infer that, because the outward manifestations of volcanic action have ceased, the internal causes of those manifestations have passed away. These are still in operation, and must make themselves felt when the internal forces have recovered their exhausted energies; but perhaps not to the same extent as before.

(f.) Region of the Snake River.--The tract of country bordering the Snake River in Idaho and Washington is remarkable for the vast sheets of plateau-basalt with which it is overspread, extending sometimes in one great flood farther than the eye can reach, and what is still more remarkable, they are often unaccompanied by any visible craters or vents of eruption. In Oregon the plateau-basalt is at least 2,000 feet in thickness, and where traversed by the Columbia River it reaches a thickness of about 3,000 feet. The Snake and Columbia rivers are lined by walls of volcanic rock, basaltic above, trachytic below, for a distance of, in the former, one hundred, in the latter, two hundred, miles. Captain Dutton, in describing the High Plateau of Utah, observes that the lavas appear to have welled up in mighty floods without any of that explosive violence generally characteristic of volcanic action. This extravasated matter has spread over wide fields, deluging the surrounding country like a tide in a bay, and overflowing all inequalities. Here also we have evidence of older volcanic cones buried beneath seas of lava subsequently extruded.

(g.) Fissures of Eruption.--The absence, or rarity, of volcanic craters or cones of eruption in the neighbourhood of these great sheets has led American geologists to the conclusion that the lavas were in many cases extruded from fissures in the earth's crust rather than from ordinary craters.[7] This view is also urged by Sir A. Geikie, who visited the Utah region of the Snake River in 1880, and has vividly described the impression produced by the sight of these vast fields of basaltic lava. He says, "We found that the older trachytic lavas of the hills had been deeply trenched by the lateral valleys, and that all these valleys had a floor of black basalt that had been poured out as the last of the molten materials from the now extinct volcanoes. There were no visible cones or vents from which these floods of basalt could have proceeded. We rode for hours by the margin of a vast plain of basalt stretching southward and westward as far as the eye could reach.... I realised the truth of an assertion made first by Richthofen,[8] that our modern volcanoes, such as Vesuvius and Etna, present us with by no means the grandest type of volcanic action, but rather belong to a time of failing activity. There have been periods of tremendous volcanic energy, when instead of escaping from a local vent, like a Vesuvian cone, the lava has found its way to the surface by innumerable fissures opened for it in the solid crust of the globe over thousands of square miles."[9]

(h.) Volcanic History of Western America.--The general succession of volcanic events throughout the region of Western America appears to have been somewhat as follows:--[10]

The earliest volcanic eruptions occurred in the later Eocene epoch and were continued into the succeeding Miocene stage. These consisted of rocks moderately rich in silica, and are grouped under the heads of propylite and andesite. To these succeeded during the Pliocene epoch still more highly silicated rocks of trachytic type, consisting of sanidine and oligoclase trachytes. Then came eruptions of rhyolite during the later Pliocene and Pleistocene epochs; and lastly, after a period of cessation, during which the rocks just described were greatly eroded, came the great eruptions of basaltic lava, deluging the plains, winding round the cones or plateaux of the older lavas, descending into the river valleys and flooding the lake beds, issuing forth from both vents and fissures, and continuing intermittently down almost into the present day--certainly into the period of man's appearance on the scene. Thus the volcanic history of Western America corresponds remarkably to that of the European regions with which we have previously dealt, both as regards the succession of the various lavas and the epochs of their eruption.

(i.) The Yellowstone Park.--The geysers and hot springs of the Yellowstone Park, like those in Iceland and New Zealand, are special manifestations of volcanic action, generally in its secondary or moribund stage. The geysers of the Yellowstone occur on a grand scale; the eruptions are frequent, and the water is projected into the air to a height of over 200 feet. Most of these are intermittent, like the remarkable one known as Old Faithful, the Castle Geyser, and the Giantess Geyser described by Dr. Hayden, which ejects the water to a height of 250 feet. The geyser-waters hold large quantities of silica and sulphur in solution, owing to their high temperature under great pressure, and these minerals are precipitated upon the cooling of the waters in the air, and form circular basins, often gorgeously tinted with red and yellow colours.[11]

[1] J. W. Powell, Exploration of the Caverns of the Colorado, pp. 114, 196. Major Powell describes a fault or fissure through which floods of lava have been forced up from beneath and have been poured over the surface. Many cinder-cones are planted along the line of this fissure.

[2] Capt. C. E. Dutton. Sixth Ann. Rep. U.S. Geol. Survey, 1884-85.

[3] Dutton, loc. cit., chap. iv. p. 165.

[4] Amer. Jour. Science, vol. 3., ser. (1871). A beautiful map of this mountain is given in the Fifth Annual Report, U.S. Geol. Survey, 1883-84. Plate 44.

[5] Daubeny, loc. cit., p. 474.

[6] Gilbert, Monograph U.S. Geol. Survey, vol. i. (1890).

[7] Powell, Exploration of the Colorado River, p. 177, etc. (1875). Hayden, Rep. U.S. Geol. Survey of the Colorado, etc. (1871-80).

[8] Richthofen, Natural System of Volcanic Rocks, Mem. California Acad. Sciences, vol. i. (1868).

[9] Geikie, Geological Sketches at Home and Abroad, p. 271 (1882).

[10] Prestwich, Geology, vol. i. p. 370, quoting from Richthofen.

[11] The origin of geysers is variously explained; see Prestwich, Geology, vol. i. p. 170. They are probably due to heated waters suddenly converted into steam by contact with rock at a high temperature.

CHAPTER III.

VOLCANOES OF NEW ZEALAND.

One other region of volcanic action remains to be noticed before passing on to the consideration of those of less recent age. New Zealand is an island wherein seem to be concentrated all the phenomena of volcanic action of past and present time. Though it is doubtful if the term "active," in its full sense, can be applied to any of the existing craters (with two or three exceptions, such as Tongariro and Whakari Island), we find craters and cones in great numbers in perfectly fresh condition, extensive sheets of trachytic and basaltic lavas, ashes, and agglomerates; lava-floods descending from the ruptured craters of ashes and scori? old crater-basins converted into lakes; geysers, hot springs and fumaroles which may be counted by hundreds, and cataracts breaking over barriers of siliceous sinter; and, lastly, lofty volcanic mountains vying in magnitude with Vesuvius and Etna. All these wonderful exhibitions of moribund volcanic action seem to be concentrated in the northern island of Auckland. The southern island, which is the larger, also has its natural attractions, but they are of a different kind; chief of all is the grand range of mountains called, not inappropriately, the "Southern Alps," vying with its European representative in the loftiness of its peaks and the splendour of its snowfields and glaciers, but formed of more ancient and solid rocks than those of the northern island.

(a.) Auckland District.--We are indebted to several naturalists for our knowledge of the volcanic regions of New Zealand, but chiefly to Ferdinand von Hochstetter, whose beautiful maps and graphic descriptions leave nothing to be desired.[1] In this work Hochstetter was assisted by Julius Haast and Sir J. Hector. From their account we learn that the Isthmus of Auckland is one of the most remarkable volcanic districts in the world. It is characterised by a large number of extinct cinder-cones, in a greater or less perfect state of preservation, and giving origin to lava-streams which have poured down the sides of the hills on to the plains. Besides these are others formed of stratified tuff, with interior craters, surrounding in mural cliffs eruptive cones of scori? ashes, and lapilli; these cones are scattered over the isthmus and shores of Waitemata and Manukau. The tuff cones and craters rise from a floor of Tertiary sandstone and shale, the horizontal strata of which are laid open in the precipitous bluffs of Waitemata and Manukau harbours; they sometimes contain fossil shells of the genera Pecten, Nucula, Cardium, Turbo, and Nerit? As the volcanic tuff-beds are intermingled with the Upper Tertiary strata, it is inferred that the first outbursts of volcanic forces occurred when the region was still beneath the waters of the ocean. Cross-sections show that the different layers slope both outwards (parallel to the sides) and inwards towards the bottom of the craters. Sometimes these craters have been converted into lakes, as in the case of those of the Eifel; but generally they are dry or have a floor of morass. Of the crater-lakes, those of Kohuora, five in number, are perhaps the most remarkable; and in the case of two of these the central cones of slag appear as islets rising from the surface of the waters. The fresh-water lake Pupuka has a depth of twenty-eight fathoms. To the north of Auckland Harbour rises out of the waters of the Hauraki Gulf the cone of Rangitoto, 920 feet high, the flanks formed of rugged streams of basalt, and the summit crowned by a circular crater of slag and ash, out of the centre of which rises a second cone with the vent of eruption. This is the largest and newest of the Auckland volcanoes, and appears to have been built up by successive outpourings of basaltic lava from the central orifice, after the general elevation of the island.

[Illustration: Fig. 26.--Forms of volcanic tuff cones, with their cross-sections, in the Province of Auckland.--No. 1. Simple tuff cone with central crater; No. 2. Outer tuff cone with interior cinder cone and crater; No. 3. The same with lava-stream issuing from the interior cone.--(After Hochstetter.)]

Before leaving the description of the tuff-cones, which are a peculiar feature in the volcanic phenomena of New Zealand, and are of many forms and varieties, we must refer to that of Mount Wellington (Maunga Rei). This is a compound volcano, in which the oldest and smallest of the group is a tuff-crater-cone, exhibiting very beautifully the outward slope of its beds. Within this crater arise two cones of cinders, each with small craters. It would appear that after a long interval the larger of the two principal cones, formed of cinders and known as Mount Wellington, burst forth from the southern margin of the older tuff-cone, and, being built up to a height of 850 feet, gradually overspread the sides of its older neighbour. Mount Wellington itself has three craters, and from these large streams of basaltic lava have issued forth in a westerly direction, while a branch entered and partially filled the old tuff-crater to the northwards.

Southwards from Manukau Harbour, and extending a short distance from the coast-line to Taranaki Point, there occurs a plateau of basalt-conglomerate (Basaltkonglomerat), with sheets of basaltic lava overspreading the Tertiary strata. These plateau-basalts are intersected by eruptive masses in the form of dykes, but still there are no craters or cones of eruption to be seen; so that we may infer that the sheets, at least, were extruded from fissures in the manner of those of the Colorado or Idaho regions of America. Proceeding still further south into the interior of the island, we here find a lofty plateau of an average elevation of 2,000 feet, interposed between the Tertiary beds of the Upper and Middle Waikato, and formed of trachytic and pitch-stone tuff, amongst which arise old extinct volcanic cones, such as those of Karioi, Pirongia, Kakepuku, Maunga Tautari, Aroha, and many others. These trachytic lavas would seem to be more ancient than the basaltic, previously described.

(b.) Taupo Lake, and surrounding district.--But of all these volcanic districts, none is more remarkable than that surrounding the Taupo Lake, which lies amidst the Tertiary strata of the Upper Waikato Basin. The surface of this lake is 1,250 feet above that of the ocean, and its margin is enclosed within a border of rhyolite and pitchstone--rising into a mass of the same material 1,800 feet high on the eastern side. The form of the lake does not suggest that it is itself the crater of a volcano, but rather that it was originated by subsidence. On all sides, however, trachytic cones arise, of which the most remarkable group lies to the south of the lake, just in front of the two giant trachytic cones, the loftiest in New Zealand, one called Tongariro, rising about 6,500 feet, and the other Ruapahu, which attains an elevation of over 9,000 feet, with the summit capped by snow. These two lofty cones, standing side by side, are supposed by the Maoris to be the husband and wife to whom were born the group of smaller cones above referred to as occupying the southern shore of Taupo Lake. The volcano of Tongariro may still be considered as in a state of activity, as its two craters (Ngauruhoe and Ketetahi) constantly emit steam, and several solfataras break out on its flanks.[2]

(c.) Roto Mahana.--In a northerly direction from Tongariro, and distant from the coast by a few miles, lies in the Bay of Plenty the second of the active volcanoes of New Zealand, the volcanic island of Whakari (White Island), from the crater of which are constantly erupted vast masses of steam clouds. The distance between these two active craters is 120 nautical miles; and along the tract joining them steam-jets and geysers issue forth from the deep fissures through which the lava sheets have formerly been extruded. Numerous lakes also occupy the larger cavities in the ground; and hot-springs, steam-fumaroles and solfataras burst out in great numbers along the banks of the Roto Mahana Lake and the Kaiwaka River by which it is drained. Amongst such eruptions of hot-water and steam we might expect the formation of siliceous sinter, and the deposition of sulphur and other minerals; nor will our expectations be disappointed. For here we have the wonderful terraces of siliceous sinter deposited by the waters entering Roto Mahana as they descend from the numerous hot-springs or pools near its margin. All travellers concur in describing these terraces as the most wonderful of all the wonders of the Lake district of New Zealand--so great is their extent, and so rich and varied is their colouring.

The beautiful map of Roto Mahana on an enlarged scale by Hochstetter shows no fewer than ten large sinter terraces descending towards the margin of this lake, besides several mud-springs, fumaroles, and solfataras. But the largest and most celebrated of all the sinter terraces has within the last few years been buried from view beneath a flood of volcanic trass, or mud, an event which was as unexpected as it was unwelcome. In May, 1887, the mountain of Tarawera, which rises to the north-east of Roto Mahana, and on the line of eruption above described, suddenly burst forth into violent activity, covering the country for miles around with clouds of ashes, and, pouring down torrents of mud, completely enveloped the beautiful terrace of sinter which had previously been one of the wonders of New Zealand. By the same eruption several human beings were entombed, and their residences destroyed.

The waters of Roto Mahana, together with the hot-springs and fountains are fed from rain, and from the waters of Taupo Lake, which, sinking through fissures in the ground, come in contact with the interior heated matter, and thus steam at high temperature and pressure is generated.[3]

(d.) Moribund condition of New Zealand Volcanoes.--From what has been said, it will be inferred that in the case of New Zealand, as in those of Auvergne, the Eifel and Lower Rhine, Arabia, and Western America, we have an example of a region wherein the volcanic forces are well-nigh spent, but in which they were in a state of extraordinary activity throughout the later Tertiary, down to the commencement of the present epoch. In most of these cases the secondary phenomena of vulcanicity are abundantly manifest; but the great exhibitions of igneous action, when the plains were devastated by sheets of lava, and cones and craters were piled up through hundreds and thousands of feet, have for the present, at least, passed away.

[1] Geol.-topographischer Atlas von Neu-Seeland, von Dr. Ferd. von Hochstetter und Dr. A. Petermann. Gotha: Justus Perthes (1863). Also New Zealand, trans. by E. Sauter, Stuttgart (1867).

[2] Tongariro was visited in 1851 by Mr. H. Dyson, who describes the eruption of steam.

[3] Mr. Froude figures and describes the two terraces, the "White" and "Pink," in Oceana, 2nd edition, pp. 285-291.

PART IV.

TERTIARY VOLCANIC DISTRICTS OF THE BRITISH ISLES.

CHAPTER I.

ANTRIM.

It is an easy transition to pass from the consideration of European and other dormant, or extinct, volcanic regions to those of the British Isles, though the volcanic forces may have become in this latter instance quiescent for a somewhat longer period. In all the cases we have been considering, whether those of Central Italy, of the Rhine and Moselle, of Auvergne, or of Syria and Arabia, the cones and craters of eruption are generally present entire, or but slightly modified in form and size by the effects of time. But in the case of the Tertiary volcanic districts of the British Isles this is not so. On the contrary, these more prominent features of vulcanicity over the surface of the ground have been removed by the agents of denudation, and our observations are confined to the phenomena presented by extensive sheets of lava and beds of ash, or the stumps and necks of former vents of eruption, together with dykes of trap by which the plateau-lavas are everywhere traversed or intersected.

The volcanic region of the British Isles extends at intervals from the North-east of Ireland through the Island of Mull and adjoining districts on the mainland of Morvern and Ardnamurchan into the Isle of Skye, and comprises several smaller islets; the whole being included in the general name of the Inner Hebrides. It is doubtful if the volcanic lavas of Co. Antrim were ever physically connected with those of the west of Scotland, though they may be considered as contemporary with them; and in all cases the existing tracts of volcanic rock are mere fragments of those originally formed by the extrusion of lavas from vents of eruption. In addition to these, there are large areas of volcanic rock overspread by the waters of the ocean.

(a.) Geological Age.--The British volcanic eruptions now under consideration are all later than the Cretaceous period. Throughout Antrim, and in parts of Mull, the lavas are found resting on highly eroded faces either of the Upper Chalk (Fig. 27), or, where it has been altogether denuded away, on still older Mesozoic strata. From the relations of the basaltic sheets of Antrim to the Upper Chalk, it is clear that the latter formation, after its deposition beneath the waters of the Cretaceous seas, was elevated into dry land and exposed to a long period of subaërial erosion before the first sheets of lava invaded the surface of the ground. We are, therefore, tolerably safe in considering the first eruptions to belong to the Tertiary period; but the evidence, derived as it is exclusively from plant remains, is somewhat conflicting as to the precise epoch to which the lavas and beds of tuff containing the plant-remains are to be referred. The probabilities appear to be that they are of Miocene age; and if so, the trachytic lavas, which in Antrim are older than those containing plants, may be referred to a still earlier epoch--namely, that of the Eocene.[1] As plant remains are not very distinctive, the question regarding the exact time of the first volcanic eruptions will probably remain for ever undecided; but we are not likely to be much in error if we consider the entire volcanic period to range from the close of the Eocene to that of the Miocene; by far the greater mass of the volcanic rocks being referable to the latter epoch.

In describing the British volcanic districts it will be most convenient to deal with them in three divisions--viz., those of Antrim, Mull, and Skye, commencing with Antrim.[2]

(b.) Volcanic Area.--The great sheets of basalt and other volcanic products of the North-east of Ireland overspread almost the whole of the County Antrim, and adjoining districts of Londonderry and Tyrone, breaking off in a fine mural escarpment along the northern shore of Belfast Lough and the sea coast throughout the whole of its range from Larne Harbour to Lough Foyle; the only direction in which these features subside into the general level of the country being around the shores of Lough Neagh. Several outliers of the volcanic sheets are to be found at intervals around the great central plateau; such as those of Rathlin Island, Island Magee, and Scrabo Hill in Co. Down. The area of the basaltic plateau may be roughly estimated at 2,000 square miles.

The truncated edges of this marginal escarpment rising to levels of 1,000 to 1,260 feet, as in the case of Benevenagh in Co. Derry, and 1,825 feet at Mullaghmore, attest an originally greatly more extended range of the basaltic sheets; and it is not improbable that at the close of the Miocene epoch they extended right across the present estuary of Lough Foyle to the flanks of the mountains of Inishowen in Donegal in one direction, and to those of Slieve Croob in the other. In the direction of Scotland the promontories of Kintyre and Islay doubtless formed a part of the original margin. Throughout this vast area the volcanic lavas rest on an exceedingly varied rocky floor, both as regards composition and geological age. (See Fig. 28.) Throughout the central, southern, eastern, and northern parts of their extent, the Chalk formation may be considered to form this floor; but in the direction of Armagh and Tyrone, towards the southwestern margin, the basaltic sheets are found resting indiscriminately on Silurian, Carboniferous, and Triassic strata. The general relations of the plateau-basalts to the underlying formations show, that at the close of the Cretaceous period there had been considerable terrestrial disturbances and great subaerial denudation, resulting in some cases in the complete destruction of the whole of the Cretaceous strata, before the lava floods were poured out; owing to which, these latter are found resting on formations of older date than the Cretaceous.[3]

[1] Mr. J. Starkie Gardner, from a recent comparison of the plant-remains of Antrim and Mull, concludes that "that they might belong to any age between the beginning and the end of the warmer Eocene period; and that they cannot be of earlier, and are unlikely to be of later, date."--Trans. Palnt. Soc., vol. xxxvii. (1883).

[2] Having dealt with this district rather fully in The Physical Geology and Geography of Ireland (Edit. 1891, p. 81), and also in my Presidential Address (Section C.) at the meeting of the British Association, 1874, a brief review of the subject will be sufficient here, the reader being referred to the former treatises for fuller details. The following should also be consulted: Gen. Portlock, Geology of Londonderry and Tyrone (1843); Sir A. Geikie, "History of

Volcanic Action during the Tertiary Period in the British Isles," Trans. Roy. Soc. Edinburgh, 1888; and the Descriptive Memoirs of the Geological Survey relating to this tract of country.

[3] Owing to the superposition of the basaltic masses on beds of chalk throughout a long line of coast, we are presented with the curious spectacle of the whitest rocks in nature overlain by the blackest, as may be seen in the cliffs at Larne, Glenarm, Kinbane and Portrush. (See Fig. 27.)

CHAPTER II.

SUCCESSION OF VOLCANIC ERUPTIONS.

(c.) First Stage.--The earliest eruptions of lava in the North-east of Ireland belonged to the highly acid varieties, consisting of quartz-trachyte with tridymite.[1] This rock rises to the surface at Tardree and Brown Dod hills and Templepatrick. It consists of a light-greyish felsitic paste enclosing grains of smoke-quartz, crystals of sanidine, plagioclase and biotite, with a little magnetite and apatite. It is a rock of peculiar interest from the fact that it is almost unique in the British Islands, and has its petrological counterpart rather amongst the volcanic hills of the Siebengebirge than elsewhere. It is generally consolidated with the columnar structure.

The trachyte appears to have been extruded from one or more vents in a viscous condition, the principal vent being probably situated under Tardree mountain, where the rock occurs in greatest mass, and it probably arose as a dome-shaped mass, with a somewhat extended margin, above the floor of Chalk which formed the surface of the ground.[2] (Fig. 27.) At Templepatrick the columnar trachyte may be observed resting on the Chalk, or upon a layer of flint gravel interposed between the two rocks, and which has been thrust out of position by a later intrusion of basalt coming in from the side.[3] It is to be observed, however, that the trachytic lavas nowhere appear cropping out along with the sheets of basalt around the escarpments overlooking the sea, or inland; showing that they did not spread very far from their vents of eruption; a fact illustrating the lower viscosity, or fluidity, of the acid lavas as compared with those of the basic type.

(d.) Second Stage.--After an interval, probably of long duration, a second eruption of volcanic matter took place over the entire area; but now the acid lavas of the first stage are replaced by basic lavas. Now, for the first time, vast masses of basalt and dolerite are extruded both from vents of eruption and fissures; and, owing to their extreme viscosity, spread themselves far and wide until they reach the margin of some uprising ground of old Paleozoic or Metamorphic rocks by which the volcanic plain is almost surrounded. The great lava sheets thus produced are generally more or less amorphous, vesicular and amygdaloidal, often exhibiting the globular concentric structure, and weathering rapidly to a kind of ferruginous sand or clay under the influence of the atmosphere. Successive extrusions of these lavas produce successive beds, which are piled one over the other in some places to a depth of 600 feet; and at the close of the stage, when the volcanic forces had for the time exhausted themselves, the whole of the North-east of Ireland must have presented an aspect not unlike that of one of those great tracts of similar lava in the region of Idaho and the Snake River in Western America, described in a previous chapter.

(e.) Third Stage (Inter-volcanic).--The third stage may be described as inter-volcanic. Owing to the formation of a basin, probably not deep, and with gently sloping sides, a large lake was formed over the centre of the area above described. Its floor was basalt, and the streams from the surrounding uplands carried down leaves and stems of trees, strewing them over its bed. Occasionally eruptions of ash took place from small vents, forming the ash-beds with plants found at Ballypallidy, Glenarm, and along the coast as at Carrick-a-raide. The streams also brought down sand and gravel from the uprising domes of trachyte, and deposited them over the lake-bed along with the erupted ashes.[4] The epoch we are now referring to was one of economic importance; as, towards its close, there was an extensive deposition of pisolitic iron-ore over the floor of the lake, sometimes to the depth of two or three feet. This ore has been extensively worked in recent years.

(f.) Fourth Stage (Volcanic).--The last stage described was brought to a termination by a second outburst of basic lavas on a scale probably even grander than the preceding. These lavas consisting of basalt and dolerite, with their varieties, and extruded from vents and fissures, spread themselves in all directions over the pre-existing lake deposits or the older sheets of augitic lava, and probably entirely buried the trachytic hills. These later sheets solidified into more solid masses than those of the second stage. They form successive terraces with columnar structure, each terrace differing from that above and below it in the size and length of the columns, and separated by thin bands of "bole" (decomposed lava), often reddish in colour, clearly defining the limits of the successive lava-flows. Nowhere throughout the entire volcanic area are these successive terraces so finely laid open to view as along the north coast of Antrim, where the lofty mural cliffs, worn back into successive bays with intervening headlands by the irresistible force of the Atlantic waves, present to the spectator a vertical section from 300 to 400 feet in height, in which the successive tiers of columnar basalt, separated by thin bands of bole, are seen to rise one above the other from the water's edge to the summit of the cliff, as shown in Fig. 30. Here, also, at the western extremity of the line of cliffs we find that remarkable group of vertical basaltic columns, stretching from the base of the cliff into the Atlantic, and known far and wide by the name of "The Giant's Causeway," the upper ends of the columns forming a tolerably level surface, gently sloping seawards, and having very much the aspect of an artificial tesselated pavement on a huge scale. A portion of the Causeway, with the cliff in the background, is shown in the figure (Fig. 31). The columns are remarkable for their symmetry, being generally hexagonal, though occasionally they are pentagons, and each column is horizontally traversed by joints of the ball-and-socket form, thus dividing them into distinct courses of natural masonry. These are very well shown in the accompanying view of the remarkable basaltic pillars known as "The Chimneys," which stand up from the margin of the headland adjoining the Causeway, monuments of past denudation, as they originally formed individuals amongst the group belonging to one of the terraces in the adjoining coast.[5] (Fig. 32).

(g.) Original Thickness of the Antrim Lavas.--It is impossible to determine with certainty what may have been the original thickness of the accumulated sheets of basic lavas with their associated beds of ash and bole. The greatest known thickness of the lower zone of lavas is, as I have already stated, about 600 feet. The intermediate beds of ash and bole sometimes attain a thickness of 40 feet, and the upper group of basalt about 400 feet; these together would constitute a series of over 1,000 feet in thickness. But this amount, great as it is, is undoubtedly below the original maximum, as the uppermost sheets have been removed by denuding agencies, we know not to what extent. Nor is it of any great importance. Sufficient remains to enable us to form a just conception of the magnitude both as regards thickness and extent of the erupted matter of the Miocene period over the North-east of Ireland and adjoining submerged tracts, and of the magnitude of the volcanic operations necessary for the production of such masses.

(h.) Volcanic Necks.--As already remarked, no craters of eruption survive throughout the volcanic region of the North-east of Ireland, owing to the enormous extent of the denudation which this region has undergone since the Miocene Epoch; but the old "necks" of such craters--in other words, the pipes filled with either solid basalt, or basalt and ashes--are still to be found at intervals over the whole area. Owing to the greater solidity of the lava which filled up these "necks" over the plateau-basaltic sheets which surround them, they appear as bosses or hills rising above the general level of the ground. One of these bosses of highly columnar basalt occurs between Portrush and Bushmills, not far from Dunluce Castle, another at Scawt Hill, near Glenarm, and a third at Carmoney Hill above Belfast Lough. But by far the most prominent of these old solidified vents of eruption is that of Sleamish, a conspicuous mountain which rises above the general level of the plateau near Ballymena, and attains an elevation of 1,437 feet above the sea. Seen from the west, the mountain has the appearance of a round-topped cone; but on examination it is found to be in reality a huge dyke, breaking off abruptly towards the north-west, in which direction it reaches its greatest height, then sloping downwards towards the east. This form suggests that Sleamish is in reality one of the fissure-vents of eruption rather than the neck of an old volcano. The rock of which it is formed consists of exceedingly massive, coarsely-crystalline dolerite, rich in olivine, and divided into large quadrangular blocks by parallel joint planes. Its junction with the plateau-basalt from which it rises can nowhere be seen; but at the nearest point where the two rocks are traceable the plateau-basalt appears to be somewhat indurated; breaking with a splintery fracture and a sharp ring under the hammer, suggesting that the lava of Sleamish had been extruded through the horizontal sheets, and had considerably indurated the portions in contact with, or in proximity to, it.[6] Amongst the vents filled with ash and agglomerate, the most remarkable is that of Carrick-a-raide, near Ballycastle. It forms this rocky island and a portion of the adjoining coast, where the beds of ash are finely displayed; consisting of fragments and bombs of basalt, with pieces of chalk, flint, and peperino, which is irregularly bedded. These ash-beds attain a thickness of about 120 feet just below the road to Ballycastle, but rapidly tail out in both directions from the locality of the vent. Just below the ash-beds, the white chalk with flints may be seen extending down into the sea-bed. Nowhere in Antrim is there such a display of volcanic ash and agglomerate as at this spot.[7]

(i.) Dykes: Conditions under which they were Erupted.--No one can visit the geological sections in Co. Antrim and the adjoining districts of Down, Armagh, Derry, and Tyrone, without being struck by the great number and variety of the igneous dykes by which the rocks are traversed. The great majority of these dykes are basaltic, and they are found traversing all the formations, including the Cretaceous and Tertiary basaltic sheets. The Carlingford and Mourne Mountains are seamed with such dykes, and they are splendidly laid open to view along the coast south of Newcastle in Co. Down, as also along the Antrim coast from Belfast to Larne. The fine old castle of Carrickfergus has its foundations on one of those dyke-like intrusions, but one of greater size than ordinary. All the dykes here referred to are not, however, of the same age, as is conclusively proved by sections amongst the Mourne Mountains where cliffs of Lower Silurian strata, superimposed on the intrusive granite of the district, exhibit two sets of basaltic dykes--one (the older) abruptly terminated at the granite margin, the other and newer penetrating the granite and Silurian rocks alike. It is not improbable that the older dykes belong to the Carboniferous or Permian age, while the newer are with equal probability of Tertiary age. Sir A. Geikie has shown that the Tertiary dykes of the North of Ireland are representatives of others occurring at intervals over the North of England, and Central and Western Scotland, all pointing towards the central region of volcanic activity; or in a parallel direction thereto, approximating to the N.W. in Ireland, the Island of Islay, and East Argyleshire, but in the centre of Scotland generally ranging from east to west.[8] The area affected by the dykes of undoubted Tertiary age Geikie estimates at no less than 40,000 square miles--a territory greater than either Scotland or Ireland, and equal to more than a third of the total land-surface of the British Isles;[9] and he regards them as posterior "to the rest of the geological structures of the regions which they traverse." It is clear that the dykes referred to belong to one great system of eruption or intrusion; and they may be regarded as the manifestation of the final effort of internal forces over this region of the British Isles. They testify to the existence of a continuous magma (or shell) of augitic lava beneath the crust; and as the aggregate horizontal extent of all these dykes, or of the fissures which they fill, must be very considerable, it is clear that the crust through which they have been extruded has received an accession of horizontal space, and has been fissured by forces acting from beneath, as the late Mr. Hopkins, of Cambridge, had explained on mechanical grounds in his elaborate essay many years ago.[10] This view occurred to myself when examining the region of the North-east of Ireland, but I was not then aware that it had been dealt with on mathematical principles by so eminent a mathematician. The bulging of the crust is a necessary consequence of the absence of plication of the strata due to the extrusion of this enormous quantity of molten lava; and the intrusion of thousands of dykes over the North-east of Ireland, unaccompanied by foldings of the strata, must have added a horizontal space of several thousand feet to that region.[11]

[1] A peculiar form of crystalline quartz first recognized in this rock by a distinguished German petrologist, the late Prof. A. von Lasaulx, who visited the district in 1876.

[2] Sir A. Geikie has disputed the correctness of the view, which I advocated as far back as 1874, that the trachytic lavas of Antrim are the earliest products of volcanic action; but at the time he wrote his paper on the volcanic history of these islands, it was not known that pebbles of this trachyte are largely distributed amongst the ash-beds which occur in the very midst of the overlying basaltic sheets, as I shall have to explain later on. This discovery puts the question at rest as regards the relations of the two sets of rocks.

[3] This remarkable section at the chalk quarries of Templepatrick the author has figured and described in the Physical Geology and Geography of Ireland, p. 99, 2nd edit. (1891), where the reader will find the subject discussed more fully than can be done here.

[4] These pebbles were first noticed by Mr. McHenry, of the Irish Geological Survey, in 1890.

[5] The vertical position of the columns of the Giant's Causeway is rather enigmatical. The Causeway cannot be a dyke, as has often been supposed, otherwise the columns would have been horizontal, i.e., at right angles to the sides of the dyke. Mr. R. G. Symes, of the Geological Survey, has suggested

that the Causeway columns have been vertically lowered between two lines of fault, and that originally they formed a portion of the tier of beautiful columns seen in the cliff above, and known as "The Organ."

[6] Sleamish and several other of the Antrim vents are described by Sir A. Geikie in the monograph already referred to, loc. cit., p. 101, et seq. Also in the Expl. Memoirs of the Geological Survey of Ireland.

[7] A diagrammatised section of the Carrick-a-raide volcanic neck is given by Sir A. Geikie, loc. cit., p. 105.

[8] Geikie, loc. cit., p. 29, et seq.

[9] P. 32. The view that the crust of the earth has been horizontally extended by the intrusion of dykes is noticed by McCulloch in reference to the dykes of Skye.

[10] Hopkins, Cambridge Phil. Trans., vol. vi. p. 1 (1836).

[11] As suggested in my Presidential Address to Section C. of the British Association at Belfast, 1874.

CHAPTER III.

ISLAND OF MULL AND ADJOINING COAST.

The Island of Mull, with the adjoining districts of Morvern and Ardnamurchan, forms the more southern of the two chief centres of Tertiary volcanic eruptions in the West of Scotland, that of Skye being the more northern. These districts have been the subject of critical and detailed study by several geologists, from McCulloch down to the present day; and amongst the more recent, Sir Archibald Geikie and Professor Judd hold the chief place. Unfortunately, the interpretation of the volcanic phenomena by these two accomplished observers has led them to very different conclusions as regards several important points in the volcanic history of these groups of islands; as, for example, regarding the relative ages of the plateau-basalts and the acid rocks, such as the trachytes and granophyres; again as regards the presence of distinct centres of eruption; and also as regards the relations of the gabbros of Skye to the basaltic sheets. Such being the case, it would appear the height of rashness on the part of the writer, especially in the absence of a detailed examination of the sections over the whole region, to venture on a statement of opinion regarding the points at issue; and he must, therefore, content himself with a brief account of the phenomena as gathered from a perusal of the writings of these and other observers,[1] guided also to some extent by the analogous phenomena presented by the volcanic region of the North-east of Ireland.

(a.) General Features.--As in the case of the Antrim district, the Island of Mull and adjoining tracts present us with the spectacle of a vast accumulation of basaltic lava-flows, piled layer upon layer, with intervening beds of bole and tuff, up to a thickness, according to Geikie, of about 3,500 feet. At the grand headland of Gribon, on the west coast, the basaltic sheets are seen to rise in one sheer sweep to a height of 1,600 feet, and then to stretch away with a slight easterly dip under Ben More at a distance of some eight miles. This mountain, the upper part of which is formed of beds of ashes, reaches an elevation of 3,169 feet, so that the accumulated thickness of the beds of basalt under the higher part of the mountain must be at least equal to the amount stated above--that is, twice as great as the representative masses of Antrim. The base of the volcanic series is seen at Carsaig and Gribon to rest on Cretaceous and Jurassic rocks, like those of Antrim; hence the Tertiary age is fully established by the evidence of superposition. This was further confirmed by the discovery by the Duke of Argyll,[2] some years ago (1850), of bands of flint-gravel and tuff, with dicotyledonous leaves amongst the basalts of Ardtun Head. The basement beds of tuff and gravel contain, besides pebbles of flint and chalk, others of sanidine trachyte, showing that highly acid lavas had been extruded and consolidated before the first eruption of the plateau-basalts; another point of analogy between the volcanic phenomenon of Antrim and the Inner Hebrides. These great sheets of augitic lava extend over the whole of the northern tract of Mull, the Isles of Ulva and Staffa, and for a distance of several miles inwards from the northern shore of the Sound of Mull, covering the wild moorlands of Morvern and Ardnamurchan, where they terminate in escarpments and outlying masses, indicating an originally much more extended range than at the present day. The summits of Ben More and its neighbouring height, Ben Buy, are formed of beds of ash and tuff. The volcanic plateau is, according to Judd, abruptly terminated along the southern side by a large vault, bringing the basalt in contact with Pal 鑒 zoic rocks.[3]

(b.) Granophyres.--The greater part of the tract lying to the south of Loch na Keal, which almost divides Mull into two islands, and extending southwards and eastwards to the shores of the Firth of Lorn and the Sound of Mull, is formed of a peculiar group of acid (or highly silicated) rocks, classed under the general term of "Granophyres." These rocks approach towards true granites in one direction, and through quartz-porphyry and felsite to rhyolite in another--probably depending upon the conditions of cooling and consolidation. In their mode of weathering and general appearance on a large scale, they present a marked contrast to the basic lavas with which they are in contact from the coast of L. na Keal to that of L. Buy. The nature of this contact, whether indicating the priority of the granophyres to the plateau-basalts or otherwise, is a matter of dispute between the two observers above named; but the circumstantial account given by Sir A. Geikie,[4] accompanied by drawings of special sections showing this contact, appears to prove that the granophyre is the newer of the two masses of volcanic rock, and that it has been intruded amongst the basaltic-lavas at a late period in the volcanic history of these islands. A copy of one of these sketches is here given (Fig. 33), according to which the felsite is shown to penetrate the basaltic sheets at Alt na Searmoin in Mull; other sections seen at Cruach Torr an Lochain, and on the south side of Beinn Fada, appear to lead to similar conclusions. These rocks are penetrated by numerous basaltic dykes.

(c.) Representative Rocks of Mourne and Carlingford, Ireland.--Assuming Sir A. Geikie's view to be correct, it is possible that we may have in the granite and quartz-porphyries of Mourne and Carlingford representatives of the granites, granophyres, and other acid rocks of the later period of Mull. The granite of Mourne is peculiar in structure, and differs from the ordinary type of that rock in which the silica forms the ground mass. In the case of the granite of the Mourne Mountains, the rock consists of a crystalline granular aggregate of orthoclase, albite, smoke-quartz, and mica; it is also full of drusy cavities, in which the various minerals crystallise out in very perfect form. As far as regards direct evidence, the age of this rock can only be stated to be post-Carboniferous, and earlier than certain Tertiary basaltic dykes by which it is traversed. The granophyres of Mull are traversed by similar dykes, which are representatives of the very latest stage of volcanic action in the British Islands. The author is therefore inclined to concur with Sir A. Geikie in assigning to the granite of the Mourne Mountains, and the representative felsitic rocks of the Carlingford Mountains, a Tertiary age--in which case the analogy between the volcanic phenomena of the Inner Hebrides and of the North-east of Ireland would seem to be complete.[5]

[1] Geikie, Proc. Roy. Soc. Edinburgh (1867); Brit. Assoc. Rep. (Dundee, 1867); "Tertiary Volcanic Rocks of the British Isles," Quart. Journ. Geol. Soc., vol. xxvii. p. 279; also, "History of Volcanic Action in British Isles," Trans. Roy. Soc. Edin. (1888); Judd, "On the Ancient Volcanoes of the Highlands," etc., Quart. Journ. Geol. Soc., vol. xxx. p. 233; and Volcanoes, p. 139.

[2] Brit. Assoc. Rep. for 1850, p. 70.

[3] Judd, Quart. Jour. Geol. Soc., vol. xxx. p. 242.

[4] History of Volcanic Action, etc., loc. cit. p. 153, et seq. The "Granophyres" of Geikie come under the head of "Felsites," passing into "granite" in one direction and quartz-trachyte in another, according to Judd; the proportion of silica from 69 to 75 per cent.--Quart. Jour. Geol. Soc., vol. xxx. p. 235.

[5] This view the author has expressed in a recent edition of The Physical Geology of Ireland, p. 177 (1891).

CHAPTER IV.

ISLE OF SKYE.

This is the largest and most important of all the Tertiary volcanic districts, but owing to the extensive denudation to which, in common with other Tertiary volcanic regions of the British Isles, it has been subjected, its present limits are very restricted comparatively to its original extent. Not only is this evident from the manner in which the basaltic sheets terminate along the sea-coast in grand mural cliffs, as opposite "Macleod's Maidens," and at the entrance to Lough Bracadale on the western coast, but the evidence is, according to Sir A. Geikie, still more striking along the eastern coast; showing that the Jurassic, and other older rocks there visible, were originally buried deep under the basaltic sheets which have been stripped from off that part of the country. These great plateau-basalts occupy about three-fourths of the entire island along the western and northern areas, rising into terraced mountains over 2,000 feet in height, and are deeply furrowed by glens and arms of the sea, along which the general structure of the tableland is laid open, sometimes for leagues at a time.

It is towards the south-eastern part of the island that the most interesting and important phenomena are centred; for here we meet with representatives of the acid (or highly silicated) group of rocks, and of remarkable beds of gabbro, which have long attracted the attention of petrologists. These latter beds, throughout a considerable distance round the flanks of the Cuillin Hills, are interposed between the acid rocks and the plateau-basalts; but towards the north, on approaching Lough Sligahan, the acid rocks, consisting of granophyres, quartz-porphyries, and hornblendic-granitites, are in direct contact with the plateau-basalts; and, according to the very circumstantial account of Sir A. Geikie, are intrusive into them; not only sending veins into the basaltic sheets, but also producing a marked alteration in their structure where they approach the newer intrusive mass. Equally circumstantial is the same author's account of the relations of the granophyres to the gabbros,[1] as seen at Meall Dearg and the western border of the Cuillin Hills--where the former rock may be seen to send numerous veins into the latter. Not only is this so, but the granophyre is frequently seen to truncate, and abruptly terminate some of the basaltic dykes by which the basic sheets are traversed--as in the neighbourhood of Beinn na Dubhaic. All these phenomena strongly remind us of the conditions of similar rocks amongst the mountains of Mourne and Carlingford in Ireland; where, at Barnaveve, the syenite (or hornblendic quartz-felsite) is seen to break through the masses of olivine gabbro, and send numerous veins into this latter rock.[2]

The interpretation here briefly sketched differs widely from that arrived at by Professor Judd. The granitoid masses of the Red Mountains (Beinn Dearg) and the neighbouring heights are, in his view, the roots of the great volcano from which were erupted the various lavas; the earlier eruptions producing the acid lavas, to be followed by the gabbros, and these by the plateau-basaltic sheets, which stretch away towards the north and west into several peninsulas. Thus he holds that "the rocks of basic composition were ejected subsequently to those of the acid variety," and appeals to various sections in confirmation of this view.[3] To reconcile these views is at present impossible;

but as the controversy between these two observers is probably not yet closed, there is room for hope that the true interpretation of the relations of these rocks to each other will ere long be fully established.

[1] Geikie, loc. cit., p. 161, etc.

[2] Physical Geology of Ireland, 2nd edition, p. 174 (Fig. 21). Professor Judd has also come to the conclusion that the granite of Mourne is of Tertiary age, Quart. Jour. Geol. Soc., vol. xxx. p. 275.

[3] Judd, loc. cit., p. 254.

CHAPTER V.

THE SCUIR OF EIGG.

Amongst the more remarkable of the smaller islets are those of Eigg, Rum, Canna, and Muck, lying between Mull on the south and Skye on the north, and undoubtedly at one time physically connected together. The Island of Eigg is especially remarkable for the fact, as stated by Geikie, that here we have the one solitary case of "a true superficial stream of acid lava--that of the Scuir of Eigg."[1] (Fig. 34.) This forms a sinuous ridge, composed of pitchstone of several kinds, of over two miles in length, rising from the midst of a tableland of bedded basalt and tuff to a height of 1,289 feet above the ocean; the plateau-basalt is traversed by basaltic dykes, ranging in a N.W.-S.E. direction. But what is specially remarkable is the evidence afforded by an examination of the course of the Scuir, that it follows the channel of an ancient river-valley, which has been hollowed out in the surface of the plateau. The course of this channel is indicated by the presence of a deposit of river-gravel, which in some places forms a sort of cushion between the base of the Scuir and the side of the channel. Over this gravel-bed the viscous pitchstone-lava appears to have flowed, taking possession of the river-channel, and also of the beds of several small tributary streams which flowed into the channel of the Scuir. The recent date of the pitchstone forming this remarkable mural ridge, once occupying the bed of a river-channel, is shown by the fact that the basaltic dykes which traverse the plateau-basalts are truncated by the river-gravel, which is, therefore, more recent; and, as we have seen, the pitchstone stream is more recent than the river-gravel. But at the time when this last volcanic eruption took place, the physical geography of the whole region must have been very different from that of the present time. From the character and composition of the pebbles in the old river-bed, amongst which are Cambrian sandstone, quartzite, clay-slate, and white Jurassic limestone, Sir A. Geikie concludes that when the river was flowing, the island must have been connected with the mainland to the east where the parent masses of these pebbles are found.

[Illustration: Fig. 34.--View of the Scuir of Eigg from the east. The lower portion of the mountain is formed of bedded basalt, or dolerite with numerous dykes and veins of basalt, felstone, and pitchstone; the upper cliff, or Scuir, is composed of pitchstone of newer age, the remnant of a lava flow which once filled a river channel in the basaltic sheets. A dyke, or sheet, of porphyry is seen to be interposed between the Scuir and the basaltic sheets.-- (After Geikie.)]

Effects of Denudation.--The position of the Scuir of Eigg and its relations to the basaltic sheets show the enormous amount of denudation which these latter have undergone since the stream of pitchstone-lava filled the old river channel. The walls, or banks, of the channel have been denuded away, thus converting the pitchstone casting into a projecting wall of rock. That it originally extended outwards into the ocean to a far greater distance than at present is evident from the abruptly truncated face of the cliff; and yet this remarkable volcanic mass seems to have been, perhaps, the most recent exhibition of volcanic action to be found in the British Isles. It is perhaps, on this account, the most striking of the numerous examples exhibited throughout the West of Scotland and the North-east of Ireland of the enormous amount of denudation to which these districts have been subjected since the extinction of the volcanic fires; and this at a period to which we cannot assign a date more ancient than that of the Pliocene. Yet, let us consider for a moment to what physical vicissitudes these districts have been subjected since that epoch. Assuming, as we may with confidence, that the volcanic eruptions were subaerial, and that the tracts covered by the plateau-basalts were in the condition of dry land when the eruptions commenced, in this condition they continued in the main throughout the period of volcanic activity. But the eruptions had scarcely ceased, and the lava floods and dykes become consolidated, before the succeeding glacial epoch set in; when the snows and glaciers of the Scottish Highlands gradually descending from their original mountain heights, and spreading outwards in all directions, ultimately enveloped the whole of the region we are now considering until it was entirely concealed beneath a mantle of ice moving slowly, but irresistibly, outwards towards the Atlantic, crossing the deep channels, such as the Sound of Mull and the Minch, climbing up the sides of opposing rocks and islands until even the Outer Hebrides and the North-east of Ireland were covered by one vast mantle of ice and snow. The movement of such a body of ice over the land must have been attended with a large amount of abrasion of the rocky floor; nor have the evidences of that abrasion entirely disappeared even at the present day. We still detect the grooves and scorings on the rock-surfaces where they have been protected by a coating of boulder clay; and we still find the surface strewn with the blocks and debris of that mighty ice-flood.

But whatever may have been the amount of erosion caused by the great ice-sheet, it was chiefly confined to the more or less horizontal surface-planes. Erosion of another kind was to succeed, and to produce more lasting effects on the configuration of the surface. On the disappearance of the ice-sheet, an epoch characterised by milder conditions of climate set in. This was accompanied by subsidence and submersion of large tracts of the land during the Interglacial stage; so that the sea rose to heights of several hundred feet above the present level, and has left behind stratified gravels with shells at these elevations in protected places. During this period of depression and of subsequent re-emergence the wave-action of the Atlantic waters must have told severely on the coast and islands, wearing them into cliffs and escarpments, furrowing out channels and levelling obstructions. Such action has gone on down to the present day. The North-west of Scotland and of Ireland has been subjected throughout a very lengthened period to the wear and tear of the Atlantic billows. In the case of the former, the remarkable breakwater which nature has thrown athwart the North-west Highlands in the direction of the waves, forming the chain of islands constituting the Outer Hebrides, and composed of very tough Archean gneiss and schist, has done much to retard the inroads which the waves might otherwise have made on the Isle of Skye; while Coll and Tiree, composed of similar materials, have acted with similar beneficent effect for Mull and the adjoining coasts. But such is the tremendous power of the Atlantic billows when impelled by westerly winds, that to their agency must be mainly attributed the small size of the volcanic land-surfaces as compared with their original extent, and the formation of those grand headlands which are presented by the igneous masses of Skye, Ardnamurchan, and Mull towards the west. Rain and river action, supplemented by that of glaciers, have also had a share in eroding channels and wearing down the upper surface of the ground, with the result we at present behold in the wild and broken scenery of the Inner Hebrides and adjoining coast.

[1] Geikie, loc. cit., p. 178; also Quart. Jour. Geol. Soc., vol. xxvii. p. 303.

CHAPTER VI.

ISLE OF STAFFA.

Reference has been made to this remarkable island in a former page, but some more extended notice is desirable before leaving the region of the Inner Hebrides. Along with the islands of Pladda, Treshnish, and Blackmore, Staffa is one of the outlying volcanic islands of the group, being distant about six miles from the coast of Mull, and indicates the minimum distance to which the plateau-basaltic sheets originally extended in the direction of the old marginal lands of Tiree and Coll. The island consists of successive sheets of bedded basaltic lava, with partings of tuff, one of which of considerable thickness is shown to lie at the base of the cliff on the south-west side of the island.[1] The successive lava-sheets present great varieties of structure, like those on the north coast of Antrim; some being amorphous, others columnar, with either straight or bent columns. The lava-sheet out of which Fingal's Cave is excavated consists of vertical prisms, beautifully formed, and surmounted by an amorphous mass of the same material. At the entrance of the Boat Cave we have a somewhat similar arrangement of the columns;[2] but at the Clam-shell Cave the prisms are curved, indicating some movement in the viscous mass before they had been fully consolidated.

Fingal's Cave is called after the celebrated prince of Morvern (or Morven), a province of ancient Caledonia. He is supposed to have been the father of Ossian, the Celtic bard rendered famous by Macpherson. The cave, one of many which pierce the coast-cliffs of Western Scotland, is 227 feet in length, 166 feet in height, and 40 feet in width. On all sides regular columns of basalt, some entire, others broken, rise out of the water and support the roof. The cave is only accessible in calm weather.

[1] A drawing of this cliff is given by Geikie in the Manual of Geology (Jukes and Geikie), 3rd edition, p. 277.

[2] Prestwich, Geology, vol. i. p. 281, where a view of this cave is given.

PART V.

PRE-TERTIARY VOLCANIC ROCKS.

CHAPTER I.

THE DECCAN TRAP-SERIES OF INDIA.

The great outpourings of augitic lava of Tertiary and recent times which we have been considering appear to have been anticipated in several parts of the world, more especially in Peninsular India and in Africa, and it is desirable that we should devote a few pages to the description of these remarkable volcanic formations, as they resemble, both in their mode of occurrence and general structure, some of the great lava-floods of a more recent period we have been considering. Of the districts to be described, the first which claims our notice is the Deccan.

(a.) Extent of the Volcanic Plateau.--The volcanic plateau of the Deccan stretches from the borders of the Western Ghats and the sea-coast near Bombay inland to Amarantak, at the head of the Narbudda River (long. 82?E.), and from Belgaum (lat. 15?31' N.) to near Goona (lat. 24?30'). The vast area thus circumscribed is far from representing the original extent of the tract overspread by the lava-floods, as outlying fragments of these lavas are found as far east as long. 84?E. in one direction, and at Kattiwar and Cutch in another. The present area, however, is estimated to be not less than 200,000 square miles.[1]

(b.) Nature and Thickness of the Lava-flows.--This tract is overspread almost continuously by sheets of basaltic lava, with occasional bands of fresh-water strata containing numerous shells, figured and described by Hislop, and believed by him to be of Lower Eocene age. The lava-sheets vary considerably in character, ranging from finest compact basalt to coarsely crystalline dolerite, in which olivine is abundant. The columnar structure is not prevalent, the rock being either amorphous, or weathering into concentric shells. Volcanic ash, or bole, is frequently found separating the different lava-flows; and in the upper amygdaloidal sheets numerous secondary minerals are found, such as quartz, agate and jasper, stilbite and chlorite. The total thickness of the whole series, where complete, is about 6,000 feet, divided as follows:

1. Upper trap; with ash and inter-trappean beds 1,500 feet 2. Middle trap; sheets of basalt and ash 4,000 " 3. Lower trap; basalt with inter-trappean beds 500 " -------- 6,000 " ========

Throughout the region here described these great sheets of volcanic rock are everywhere approximately horizontal, and constitute a table-land of 3,000 to 4,000 feet in elevation, breaking off in terraced escarpments, and penetrated by deep river-valleys, of which the Narbudda is the most important. The foundation rock is sometimes metamorphic schist, or gneiss, at other times sandstone referred by Hislop to Jurassic age; and in no single instance has a volcanic crater or focus of eruption been observed. But outside the central trappean area volcanic foci are numerous, as in Cutch, the Rajhipla Hills and the Lower Narbudda valley. The original excessive fluidity of the Deccan trap is proved by the remarkable horizontality of the beds over large areas, and the extensive regions covered by very thin sheets of basalt or dolerite.

(c.) Geological Age.--As regards the geological age of this great volcanic series much uncertainty exists, owing to the absence of marine forms in the inter-trappean beds. One single species, Cardita variabilis, has been observed as occurring in these beds, and in the limestone below the base of the trap at Dudukur. The facies of the forms in this limestone is Tertiary; but there is a remarkable absence of characteristic genera. On the other hand, Mr. Blanford states that the bedded traps are seen to underlie the Eocene Tertiary strata with Nummulites in Guzerat and Cutch,[2] which would appear to determine the limit of their age in one direction. On balancing the evidence, however, it is tolerably clear that the volcanic eruptions commenced towards the close of the Cretaceous period, and continued into the commencement of the Tertiary, thus bridging over the interval between the two epochs; and since the greater sheets have been exposed throughout the whole of the Tertiary and Quarternary periods, it is not surprising if they have suffered enormously from denuding agencies, and that any craters or cones of eruption that may once have existed have disappeared.

[1] The Deccan Traps have been described by Sykes, Geol. Trans., 2nd Series, vol. iv.; also Rev. S. Hislop, "On the Geology of the Neighbourhood of Nagpur, Central India," Quart. Journ. Geol. Soc., vol. x. p. 274; and Ibid., vol. xvi. p. 154. Also, H. B. Medlicott and W. T. Blanford, Manual of the Geology of India, vol. i. (1879).

[2] Blanford, Geology of Abyssinia, p. 185.

CHAPTER II.

ABYSSINIAN TABLE-LANDS.

Another region in which the volcanic phenomena bear a remarkable analogy to those of Central India, just described, is that of Abyssinia. Nor are these tracts so widely separated that they may not be considered as portions of one great volcanic area extending from Abyssinia, through Southern Arabia, into Cutch and the Deccan, in the one direction, while the great volcanic cones of Kenia and Kilimanjaro, with their surrounding tracts of volcanic matter, may be the extreme prolongations in the other. Along this tract volcanic operations are still active in the Gulf of Aden; and cones quite unchanged in form, and evidently of very recent date, abound in many places along the coast both of Arabia and Africa. The volcanic formations of this tract are, however, much more recent than those which occupy the high plateaux of Central and Southern Abyssinia of which we are about to speak.

(a.) Physical Features.--Abyssinia forms a compact region of lofty plateaux intersected by deep valleys, interposed between the basin of the Nile on the west, and the low-lying tract bordering the Red Sea and the Indian Ocean on the east. The plateaux are deeply intersected by valleys and ravines, giving birth to streams which feed the head waters of the Blue Nile (Bahr el Arak) and the Atbara. Several fine lakes lie in the lap of the mountains, of which the Zana, or Dembia, is the largest, and next Ashangi, visited by the British army on its march to Magdala in 1868, and which, from its form and the volcanic nature of the surrounding hills, appears to occupy the hollow of an extinct crater. The table-land of Abyssinia reaches its highest elevation along the eastern and southern margin, where its average height may be 8,000 to 10,000 feet; but some peaks rise to a height of 12,000 to 15,000 feet in Shoa and Ankobar.[1]

(b.) Basaltic Lava Sheets.--An enormous area of this country seems to be composed of volcanic rocks chiefly in the form of sheets of basaltic lava, which rise into high plateaux, and break off in steep--sometimes precipitous-- mural escarpments along the sides of the valleys. These are divisible into the following series:--

(1) The Ashangi Volcanic Series.--The earliest forerunners of the more recent lavas seem to have been erupted in Jurassic times, in the form of sheets of contemporaneous basalt or dolerite amongst the Antola limestones which are of this period. But the great mass of the volcanic rocks are much more recent, and may be confidently referred to the late Cretaceous or early Tertiary epochs. Their resemblance to the great trappean series of Western India, even in minute particulars, is referred to by Mr. Blanford, who suggests the view that they belong to one and the same great series of lava-flows extruded over the surface of this part of the globe. This view is inherently probable. They consist of basalts and dolerites, generally amygdaloidal, with nodules of agate and zeolite, and are frequently coated with green-earth (chlorite). Beds of volcanic ash or breccia also frequently occur, and often contain augite crystals. At Senaf? hills of trachyte passing into claystone and basalt were observed by Mr. Blanford, but it is not clear what are their relations to the plateau-basaltic sheets.[2]

(2) Magdala Volcanic Series.--This is a more recent group of volcanic lavas, chiefly distinguished from the lower, or Ashangi, group, by the occurrence of thick beds of trachyte, usually more or less crystalline, and containing beautiful crystals of sanidine. The beds of trachyte break off in precipitous scarps, and being of great thickness and perfectly horizontal, are unusually conspicuous. Mr. Blanford says, with regard to this group, that there is a remarkable resemblance in its physical aspect to the scenery of the Deccan and the higher valleys of the Western Ghats of India, but the peculiarities of the landscape are exaggerated in Abyssinia. Many of the trachytic beds are brecciated and highly columnar; sedimentary beds are also interstratified with those of volcanic origin. The Magdala group is unconformable to that of

Ashangi in some places. A still more recent group of volcanic rocks appears to occur in the neighbourhood of Senaf? consisting of amorphous masses of trachyte, often so fine-grained and compact as to pass into claystone and to resemble sandstone. At Akub Teriki the rocks appear to be in the immediate vicinity of an ancient vent of eruption.

From what has been said, it will be apparent that Abyssinia offers volcanic phenomena of great interest for the observer. There is considerable variety in the rock masses, in their mode of distribution, and in the scenery which they produce. The extensive horizontal sheets of lava are suggestive of fissure-eruption rather than of eruption through volcanic craters; and although these may have once been in existence, denudation has left no vestiges of them at the present day. In all these respects the resemblance of the volcanic phenomena to those of Peninsular India is remarkably striking; it suggests the view that they are contemporaneous as regards the time of their eruption, and similar as regards their mode of formation.

[1] W. T. Blanford, Geology of Abyssinia, pp. 151-2.

[2] Blanford, loc. cit., p. 182.

CHAPTER III.

CAPE COLONY.

Basalt of the Plateau.--The extensive sheets of plateau-basalt forming portions of the Neuweld range and the elevated table-land of Cape Colony, may be regarded as forerunners of those just described, and possibly contemporaneous with the Ashangi volcanic series of Abyssinia. The great basaltic sheets of the Cape Colony are found capping the highest elevations of the Camderboo and Stormberg ranges, as well as overspreading immense areas of less elevated land, to an extent, according to Professor A. H. Green, of at least 120,000 square miles.[1] Amongst these sheets, innumerable dykes, and masses of solid lava which filled the old vents of eruption, are to be observed. The floor upon which the lava-floods have been poured out generally consists of the "Cave Sandstone," the uppermost of a series of deposits which had previously been laid down over the bed of an extensive lake which occupied this part of Africa during the Mesozoic period. After the deposition of this sandstone, the volcanic forces appear to have burst through the crust, and from vents and fissures great floods of augitic lava, with beds of tuff, invaded the region occupied by the waters of the lake. The lava-sheets have since undergone extensive denudation, and are intersected by valleys and depressions eroded down through them into the sandstone floor beneath; and though the precise geological period at which they were extruded must remain in doubt, it appears probable that they may be referred to that of the Trias.[2]

[1] Green. "On the Geology of the Cape Colony," Quart. Jour. Geol. Soc., vol. xliv. (1888).

[2] The district lying along the south coast of Africa is described by Andrew G. Bain, in the Trans. Geol. Soc., vol. vii. (1845); but there is little information regarding the volcanic region here referred to.

CHAPTER IV.

VOLCANIC ROCKS OF PAST GEOLOGICAL PERIODS OF THE BRITISH ISLES.

It is beyond the scope of this work to describe the volcanic rocks of pre-Tertiary times over various parts of the globe. The subject is far too large to be treated otherwise than in a distinct and separate essay. I will therefore content myself with a brief enumeration of the formations of the British Isles in which contemporaneous volcanic action has been recognised.[1]

There is little evidence of volcanic action throughout the long lapse of time extending backwards from the Cretaceous to the Triassic epochs, that is to say, throughout the Mesozoic or Secondary period, and it is not till we reach the Pal 鎏 zoic strata that evidence of volcanic action unmistakably presents itself.

Permian Period.--In Ayrshire, and in the western parts of Devonshire, beds of felspathic porphyry, felstone and ash are interstratified with strata believed to be of Permian age. In Devonshire these have only recently been recognised by Dr. Irving and the author as of Permian age, the strata consisting of beds of breccia, lying at the base of the New Red Sandstone. Those of Ayrshire have long been recognised as of the same period; as they rest unconformably on the coal measures, and consist of porphyrites, melaphyres, and tuffs of volcanic origin.

Carboniferous Period.--Volcanic rocks occur amongst the coal-measures of England and Scotland, while they are also found interbedded with the Carboniferous Limestone series in Derbyshire, Scotland, and Co. Limerick in Ireland. The rocks consist chiefly of basalt, dolerite, melaphyre and felstone.

Devonian Period.--Volcanic rocks of Devonian age occur in the South of Scotland, consisting of felstone-porphyries and melaphyres; also at Boyle, in Roscommon, and amongst the Glengariff beds near Killarney in Ireland.

Upper Silurian Period.--Volcanic rocks of this stage are only known in Ireland, on the borders of Cos. Mayo and Galway, west of Lough Mask, and at the extreme headland of the Dingle Promontory in Co. Kerry. They consist of porphyrites, felstones and tuffs, or breccias, contemporaneously erupted during the Wenlock and Ludlow stages. Around the flanks of Muilrea, beds of purple quartz-felstone with tuff are interstratified with the Upper Silurian grits and slates.

Lower Silurian Period.--Volcanic action was developed on a grand scale during the Arenig and Caradoc-Bala stages, both in Wales and the Lake district, and in the Llandeilo stage in the South of Scotland. The felspathic lavas, with their associated beds of tuff and breccia, rise into some of the grandest mountain crests of North Wales, such as those of Cader Idris, Aran Mowddwy, Arenig and Moel Wyn. A similar series is also represented in Ireland, ranging from Wicklow to Waterford, forming a double group of felstones, porphyries, breccias, and ash-beds, with dykes of basalt and dolerite. The same series again appears amidst the Lower Silurian beds of Co. Louth, near Drogheda.

Metamorphic Series presumably of Lower Silurian Age.--If, as seems highly probable, the great metamorphic series of Donegal and Derry are the representatives in time of the Lower Silurian series, some of the great sheets of felspathic and hornblendic trap which they contain are referable to this epoch. These rocks have undergone a change in structure along with the sedimentary strata of which they were originally formed, so that the sheets of (presumably) augitic lava have been converted into hornblende-rock and schist. Similar masses occur in North Mayo, south of Belderg Harbour.

Cambrian Period.--In the Pass of Llanberis, along the banks of Llyn Padarn, masses of quartz-porphyry, felsite and agglomerate, or breccia, indicate volcanic action during this stage. These rocks underlie beds of conglomerate, slate and grit of the Lower Cambrian epoch, and, as Mr. Blake has shown, are clearly of volcanic origin, and pass upwards into the sedimentary strata of the period. A similar group, first recognised by Professor Sedgwick, stretches southwards from Bangor along the southern shore of the Menai Straits. Again, we find the volcanic eruptions of this epoch at St. David's, consisting of diabasic and felsitic lava, with beds of ash; and in the centre of England, amongst the grits and slates of Charnwood Forest presumably of Cambrian age, various felstones, porphyries, and volcanic breccias are found.

Thus it will be seen that every epoch, from the earliest stage of the Cambrian to the Permian, in the British Isles, gives evidence of the existence of volcanic action; from which we may infer that the originating cause, whatever it may be, has been in operation throughout all past geological time represented by living forms. The question of the condition of our globe in Arch 鎓 n times, and earlier, is one which only can be discussed on theoretic ground, and is beyond the scope of this work.

[1] The reader is referred to Sir A. Geikie's Presidential Address to the Geological Society (1891) for the latest view of this subject.

PART VI.

SPECIAL VOLCANIC AND SEISMIC PHENOMENA.

CHAPTER I.

THE ERUPTION OF KRAKATOA IN 1883.

I propose to introduce here some account of one of the most terrible outbursts of volcanic action that have taken place in modern times; namely, the eruption of the volcano of Krakatoa (a corruption of Rakata) in the strait of Sunda, between the islands of Sumatra and Java, in the year 1883. The Malay Archipelago, of which this island once formed a member, is a region where volcanic action is constant, and where the outbursts are exceptionally violent. With the great island of Borneo as a solid, non-volcanic central core, a line of volcanic islands extends from Chedooba off the coast of Pegu through Sumatra, Java, Sumbawa, Flores, and, reaching the Moluccas, stretches northwards through the Philippines into Japan and Kamtschatka. This is probably the most active volcanic belt in the world, and the recent terrible earthquake and eruption in Japan (November, 1891) gives proof that the volcanic forces are as powerful and destructive as ever.[1]

(a.) Dormant Condition down to 1680.--Down to the year 1680, this island, although from its form and structure evidently volcanic, appears to have been in a dormant state; its sides were covered with luxuriant forests, and numerous habitations dotted its shore. But in May of that year an eruption occurred, owing to which the aspect of Krakatoa as described by Vogel was entirely changed; the surface of the island when this writer passed on his voyage to Sumatra appeared burnt up and arid, while blocks of incandescent rock were being hurled into the air from four distinct points. After this first recorded eruption the island relapsed into a state of repose, and except for a stream of molten lava which issued from the northern extremity, there was no evidence of its dangerous condition. The luxuriant vegetation of the tropics speedily re-established itself, and the volcano was generally regarded as "extinct."[2] History repeats itself; and the history of Vesuvius was repeated in the case of Krakatoa.

[Illustration: Fig. 35.--Map Of The Krakatoa Group Of Islands Before The Eruption Of August 1883 (From Admiralty Chart)]

(b.) Eruption of May, 1883.[3]--On the morning of May 20, 1883, the inhabitants of Batavia, of Buitenzorg, and neighbouring localities, were surprised by a confused noise, mingled with detonations resembling the firing of artillery. The phenomena commenced between ten and eleven o'clock in the morning, and soon acquired such intensity as to cause general alarm. The detonations were accompanied by tremblings of the ground, of buildings and various objects contained in dwellings; but it was generally admitted that these did not proceed from earthquake shocks, but from atmospheric vibrations. No deviation of the magnetic needle was observed at the Meteorological Institute of Batavia; but a vertical oscillation was apparent, and persons who listened with the ear placed on the ground, even during the most violent detonations, could hear no subterranean noise whatever. It became clear that the sounds came from some volcano burst into activity; but it is strange that for two whole days it remained uncertain what was the particular volcano to which the phenomena were to be referred. The detonations appeared, indeed, to come from the direction of Krakatoa; but from Serang, Anjer, and Merak, localities situated much nearer Krakatoa than Batavia, the telegraph announced that neither detonations nor atmospheric vibrations had been perceived. The distance between Batavia and Krakatoa is ninety-three English miles. The doubts thus experienced were, however, soon put to rest by the arrival of an American vessel under the command of A. R. Thomas, and of other ships which hailed from the straits of Sunda. From their accounts it was ascertained that in the direction of Krakatoa the heavens were clouded with ashes, and that a grand column of smoke, illumined from time to time by flashes of flame, arose from above the island. Thus after a repose of more than two hundred years, "the peaceable isle of Krakatoa, inhabited, and covered by thick forests, was suddenly awakened from its condition of fancied security."

[Illustration: Fig. 36.--Section from Verlaten Island through Krakatoa, to show the outline before and after the eruption of August, 1888. The continuous line shows the former; the dotted line and shading, the latter; from which it will be observed that the original island has to a large extent disappeared. The line of section is shown in Fig. 35.]

(c.) Form and Appearance of the Island before the Eruption of 1883.--From surveys made in 1849 and 1881, it would appear that the island of Krakatoa consisted of three mountains or groups of mountains (Figs. 35, 36); the southern formed by the cone of Rakata (properly so called), rising with a scarped face above the sea to a height of over 800 metres (2,622 feet). Adjoining this cone, and rising from the centre of the island, came the group of Danan, composed of many summits, probably forming part of the enceinte annulaire of a crater. And near the northern extremity of the isle, a third group of mammelated heights could be recognised under the general name of Perboewatan, from which issued several obsidian lava-flows, with a steep slope; these dated back perhaps to the period of the first known eruption of 1680. This large and mountainous island as it existed at the beginning of May, 1883, has been entirely destroyed by the terrible eruptions of that year, with the exception of the peripheric rim (composed of the most ancient of the volcanic rocks, andesite), of which Verlaten Island and Rakata formed a part, and one very small islet, which is noted on the maps as "rots" (rock), and on the new map of the Straits of Sunda of the Dutch Navy as that of "Bootsmansrots."[4]

As shown by the map in the Report of the Royal Society, the group of islands which existed previous to 1883 were but the unsubmerged portions of one vast volcanic crater, built up of a remarkable variety of lava allied to the andesite of the Java volcanoes, but having a larger percentage of silica, and hence falling under the head of "enstatite-dacite."[5] That these volcanic rocks are of very recent origin is shown by the fact, ascertained by Verbeek, that beneath them occur deposits of Post-Tertiary age, and that these in turn rest on the Tertiary strata which are widely distributed through Sumatra, Java, and the adjoining islands. According to the reasoning of Professor Judd, the Krakatoa group at an early period of its history presented the form of a magnificent crater-cone, several miles in circumference at the base, which subsequent eruptions shattered into fragments or blew into the air in the form of dust, ashes, and blocks of lava, while the central part collapsed and fell in, leaving a vast circular ring like the ancient crater of Somma (see Fig. 6, p. 43), and he supposes the former eruptions to have been on a scale exceeding in magnificence those which have caused such world-wide interest within the last few years.

(d.) Eruption of 26th to 28th of August.--It was, as we have seen, in the month of May that, in the language of Chev. Verbeek, "the volcano of Krakatoa chose to announce in a high voice to the inhabitants of the Archipelago that, although almost nothing amongst the many colossal volcanic mountains of the Indies, it yielded to none of them in regard to its power." These eruptions were, however, only premonitory of the tremendous and terrible explosion which was to commence on Sunday, the 26th of August, and which continued for several days subsequently. A little after noon of that day, a rumbling noise accompanied by short and feeble explosions was heard at Buitenzorg, coming from the direction of Krakatoa; and similar sounds were heard at Anjer and Batavia a little later. Soon these detonations augmented in intensity, especially about five o'clock in the evening; and news was afterwards received that the sounds had been heard in the isle of Java. These sounds increased still more during the night, so that few persons living on the west side of the isle of Java were able to sleep. At seven in the morning there came a crash so formidable, that those who had hoped for a little sleep at Buitenzorg leaped from their beds. Meanwhile the sky, which had up to this time been clear, became overcast, so that by ten o'clock it became necessary to have recourse to lamps, and the air became charged with vapour. Occasional shocks of earthquake were now felt. Darkness became general all over the straits and the bordering coasts. Showers of ashes began to fall. The repeated shocks of earthquake, and the rapid discharges of subterranean artillery, all combined to show that an eruption of even greater violence than that of May was in progress at the isle of Krakatoa.

But the most interested witnesses to this terrible outburst were those on board the ships plying through the straits. Amongst these was the Charles Bal, a British vessel under the command of Captain Watson. This ship was ten miles south of the volcano on Sunday afternoon, and therefore well in sight of the island at the time when the volcano had entered upon its paroxysmal state of action. Captain Watson describes the island as being covered by a dense black cloud, while sounds like the discharges of artillery occurred at intervals of a second of time; and a crackling noise (probably arising from the impact of fragments of rock ascending and descending in the atmosphere) was heard by those on board. These appearances became so threatening towards five o'clock in the evening, that the commander feared to continue his voyage and began to shorten sail. From five to six o'clock a rain of pumice in large pieces, quite warm, fell upon the ship, which was one of those that escaped destruction during this terrible night.[6]

(e.) Electrical Phenomena.--During this eruption, electrical phenomena of great splendour were observed. Captain Wooldbridge, viewing the eruption in the afternoon of the 26th from a distance of forty miles, speaks of a great vapour-cloud looking like an immense wall being momentarily lighted up "by bursts of forked lightning like large serpents rushing through the air. After sunset this dark wall resembled a blood-red curtain, with edges of all shades of yellow, the whole of a murky tinge, through which gleamed fierce flashes of lightning." As Professor Judd observes, the abundant generation of atmospheric electricity is a familiar phenomenon in all volcanic eruptions on a grand scale. The steam-jets rushing through the orifices of the earth's crust constitute an enormous hydro-electrical engine, and the friction of the ejected materials striking against one another in their ascent and descent also does much in the way of generating electricity.[7] It has been estimated by several observers that the column of watery vapour ascended to a height of from twelve to seventeen and even twenty-three miles; and on reaching the upper strata of the atmosphere, it spread itself out in a vast canopy resembling "the pine-tree" form of Vesuvian eruptions; and throughout the long night of the 27th this canopy continued to extend laterally, and the particles of dust which it enclosed began to descend slowly through the air.

(f.) Formation of Waves.--This tremendous outburst of volcanic forces, which to a greater or less extent influenced the entire surface of the globe, gave rise to waves which traversed both air and ocean; and in consequence of the large number of observatories scattered all over the globe, and the excellence and delicacy of the instruments of observation, we are put in possession of the remarkable results which have been obtained from the collection of the observations in the hands of competent specialists. The results are related in extenso in the Report of the Royal Society, illustrated by maps and diagrams, and are worthy of careful study by those interested in terrestrial phenomena. A brief summary is all that can be given here, but it will probably suffice to bring home to the reader the magnitude and grandeur of the eruption.

The vibrations or waves generated in August, 1883, at Krakatoa may be arranged under three heads: (1) Atmospheric Waves; (2) Sound Waves; and (3) Oceanic Waves; which I will touch upon in the order here stated.

(1) Atmospheric Waves.--These phenomena have been ably handled by General Strachey,[8] from a large number of observations extending all over the globe. From these it has been clearly established that an atmospheric wave, originating at Krakatoa as a centre, expanded outwards in a circular form and travelled onwards till it became a great circle at a distance of 180 degrees from its point of origin, after which it still advanced, but now gradually contracting to a node at the antipodes of Krakatoa; that is to say, at a point over the surface of North America, situated in lat. 6?N. and long. 72?W. (or thereabout). Having attained this position, the wave was reflected or reproduced, expanding outwards for 180 degrees and travelling backwards again to Krakatoa, from which it again started, and returning to its original form again overspread the globe. This wonderful repetition, due to the spherical form of the earth, was observed no fewer than seven times, though

with such diminished force as ultimately to be outside the range of observation by the most sensitive instruments. It is one of the triumphs of modern scientific appliances that the course of such a wave, generated in a fluid surrounding a globe, which might be demonstrated on mathematical principles, has been actually determined by experiments carried on over so great an area.

(2) Sound Waves.--If the sound-waves produced at the time of maximum eruption were not quite as far-reaching as those of the air, they were certainly sufficiently surprising to be almost incredible, were it not that they rest, both as regards time and character, upon incontestible authority. The sound of the eruption, resembling that of the discharge of artillery, was heard not only over nearly all parts of Sumatra, Java, and the coast of Borneo opposite the Straits of Sunda, but at places over two thousand miles distant from the scene of the explosions. Detailed accounts, collected with great care, are given in the Report of the Royal Society, from which the following are selected as examples:--

1. At the port of Acheen, at the northern extremity of Sumatra, distant 1,073 miles, it was supposed that the port was being attacked, and the troops were put under arms.

2. At Singapore, distant 522 miles, two steamers were dispatched to look out for the vessel which was supposed to be firing guns as signals of distress.

3. At Bankok, in Siam, distant 1,413 miles, the report was heard on the 27th; as also at Labuan, in Borneo, distant 1,037 miles.

4. At places in the Philippine Islands, distant about 1,450 miles, detonations were heard on the 27th, at the time of the eruption.

The above places lie northwards of Krakatoa. In the opposite direction, we have the following examples:--

5. At Perth, in Western Australia, distant 1,092 miles, sounds as of guns firing at sea were heard; and at the Victorian Plains, distant about 1,700 miles, similar sounds were heard.

6. In South Australia, at Alice's Springs, Undoolga, and other places at distances of over 2,000 miles, the sounds of the eruption were also heard.

7. In a westerly direction at Dutch Bay, Ceylon, distant 2,058 miles, the sounds were heard between 7 a.m. and 10 a.m. on the morning of the 27th of August.

8. Lastly, at the Chagos Islands, distant 2,267 miles, the detonations were audible between 10 and 11 a.m. of the same day.

Some of the above distances are so great that we may fail to realise them; but they will be more easily appreciated, perhaps, if we change the localities to our own side of the globe, and take two or three cases with similar distances. Then, if the eruption had taken place amongst the volcanoes of the Canaries, the detonations would have been heard at Gibraltar, at Lisbon, at Portsmouth, Southampton, Cork, and probably at Dublin and Liverpool; or, again, supposing the eruption had taken place on the coast of Iceland, the report would have been heard all over the western and northern coasts of the British Isles, as well as at Amsterdam and the Hague. The enormous distance to which the sound travelled in the case of Krakatoa was greatly due to the fact that the explosions took place at the surface of the sea, and the sound was carried along that surface uninterruptedly to the localities recorded; a range of mountains intervening would have cut off the sound-wave at a comparatively short distance from its source.

(3) Oceanic Waves.--As may be supposed, the eruption gave rise to great agitation of the ocean waters with various degrees of vertical oscillation; but according to the conclusions of Captain Wharton, founded on numerous data, the greatest wave seems to have originated at Krakatoa about 10 a.m. on the 27th of August, rising on the coasts of the Straits of Sunda to a height of fifty feet above the ordinary sea-level. This wave appears to have been observed over at least half the globe. It travelled westwards to the coast of Hindostan and Southern Arabia, ultimately reaching the coasts of France and England. Eastwards it struck the coast of Australia, New Zealand, the Sandwich Islands, Alaska, and the western coast of North America; so that it was only the continent of North and South America which formed a barrier (and that not absolute) to the circulation of this oceanic wave all over the globe. The destruction to life and property caused by this wave along the coasts of

Sunda was very great. Combined with the earthquake shocks (which, however, were not very severe), the tremendous storm of wind, the fall of ashes and cinders, and the changes in the sea-bed, it produced in the Straits of Sunda for some time after the eruption a disastrous transformation. Lighthouses had been swept away; all the old familiar landmarks on the shore were obscured by a vast deposit of volcanic dust; the sea itself was encumbered with enormous quantities of floating pumice, in many places of such thickness that no vessel could force its way through them; and for months after the eruption one of the principal channels was greatly obstructed by two islands which had arisen in its midst. The Sebesi channel was completely blocked by banks composed of volcanic materials, and two portions of these banks rose above the sea as islands, which received the name of "Steers Island" and "Calmeyer Island"; but these, by the action of the waves, have since been completely swept away, and the materials strewn over the bed of the sea.[9]

(g.) Atmospheric Effects.--But the face of nature, even in her most terrific and repulsive aspect, is seldom altogether unrelieved by some traces of beauty. In contrast to the fearful and disastrous phenomena just described, is to be placed the splendour of the heavens, witnessed all over the central regions of the globe throughout a period of several months after the eruption of 1883, which has been ably treated by the Hon. Rollo Russell and Mr. C. D. Archibald, in the Royal Society's Report.

When the particles of lava and ashes mingled with vapour were projected into the air with a velocity greater than that of a ball discharged from the largest Armstrong gun, these materials were carried by the prevalent trade-winds in a westerly direction, and some of them fell on the deck of ships sailing in the Indian Ocean as far as long. 80?E., as in the case of the British Empire--on which the particles fell on the 29th of August, at a distance of 1,600 miles from Krakatoa. But far beyond this limit, the finer particles of dust (or rather minute crystals of felspar and other minerals), mingled with vapour of water, were carried by the higher currents of the air as far as the Seychelles and Africa,--not only the East coast, but also the West, as Cape Coast Castle on the Gold Coast; to Paramaribo, Trinidad, Panama, the Sandwich Isles, Ceylon and British India, at all of which places during the month of September the sun assumed tints of blue or green, as did also the moon just before and after the appearance of the stars;[10] and from the latter end of September and for several months, the sky was remarkable for its magnificent coloration; passing from crimson through purple to yellow, and melting away in azure tints which were visible in Europe and the British Isles; while a large corona was observed round both the sun and moon. These beautiful sky effects were objects of general observation throughout the latter part of the year 1883 and commencement of the following year.

The explanation of these phenomena may be briefly stated. The fine particles, consisting for the most part of translucent crystals, or fragments of crystals, formed a canopy high up in the atmosphere, being gradually spread over both sides of the equator till it formed a broad belt, through which the rays of the sun and moon were refracted. Towards dawn and sunset they were refracted and reflected from the facets of the crystal, and thus underwent decomposition into the prismatic colours; as do the rays of the sun when refracted and reflected from the particles of moisture in a rain-cloud. The subject is one which cannot be fully dealt with here, and is rather outside the scope of this work.

(h.) Origin of the Eruption.--The ultimate cause of volcanic eruptions is treated in a subsequent chapter, nor is that of Krakatoa essentially different from others. It was remarkable, however, both for the magnitude of the forces evoked and the stupendous scale of the resulting phenomena. It is evident that water played an important part in these phenomena, though not as the prime mover;--any more than water in the boiler of a locomotive is the prime mover in the generation of the steam. Without the fuel in the furnace the steam would not be produced; and the amount of steam generated will be proportional to the quantity and heat of the fuel in the furnace and the quantity of water in the boiler. In the case of Krakatoa, both these elements were enormous and inexhaustible. The volcanic chimney (or system of chimneys), being situated on an island, was readily accessible to the waters of the ocean when fissures gave them access to the interior molten matter. That such fissures were opened we may well believe. The earthquakes which occurred at the beginning of May, and later on, on the 27th of that month, may indicate movements of the crust by which water gained access. It appears that in May the only crater in a state of activity was that of Perboewatan; in June another crater came into action, connected with Danan in the centre of the island, and in August a third burst forth. Thus there was progressive activity up to the commencement of the grand eruption of the

26th of that month.[11] During this last paroxysmal stage, the centre of the island gave way and sunk down, when the waters of the ocean gained free access, and meeting with the columns of molten matter rising from below originated the prodigious masses of steam which rose into the air.

(i.) Cause of the Detonations.--The detonations which accompanied the last great eruption are repeatedly referred to in all the accounts. These may have been due, not only to the sudden explosions of steam directly produced by the ocean water coming in contact with the molten lava, but by dissociation of the vapour of water at the critical point of temperature into its elements of oxygen and hydrogen; the reunion of these elements at the required temperature would also result in explosions.

The phenomena attending this great volcanic eruption, so carefully tabulated and critically examined, will henceforth be referred to as constituting an epoch in the history of volcanic action over the globe, and be of immense value for reference and comparison.

[1] The eruption of Krakatoa has been the subject of an elaborate Report published by the Royal Society, and is also described in a work by Chevalier R. D. M. Verbeek, Ingenieur en Chef des Mines, and published by order of the Governor-General of the Netherland Indies (1886). See also an Article by Sir R. S. Ball in the Contemporary Review for November, 1888.

[2] Verbeek, loc. cit., p. 4.

[3] The account of this eruption is a free translation from Verbeek.

[4] Verbeek, loc. cit., p. 160.

[5] Judd, Rep. R. S.

[6] A fuller account by Prof. Judd will be found in the Report of the Royal Society, p. 14. Several vessels at anchor were driven ashore on the straits owing to the strong wind which arose.

[7] Judd, Report, p. 21.

[8] Report, Part ii.

[9] In this eruption, 36,380 human beings perished, of whom 37 were Europeans; 163 villages (kampoengs) were entirely, and 132 partially, destroyed.--Verbeek, loc. cit., p. 79.

[10] Verbeek, loc. cit., p. 144-5. The dust put a girdle round the earth in thirteen days.

[11] Verbeek, loc. cit., p. 30.

CHAPTER II.

EARTHQUAKES.

Connection of Earthquakes with Volcanic Action.--The connection between earthquake shocks and volcanic eruptions is now so generally recognised that it is unnecessary to insist upon it here. All volcanic districts over the globe are specially liable to vibrations of the crust; but at the same time it is to be recollected that these movements visit countries occasionally from which volcanoes, either recent or extinct, are absent; in which cases we may consider earthquake shocks to be abortive attempts to originate volcanic action.

(a.) Origin.--From the numerous observations which have been made regarding the nature of these phenomena by Hopkins, Lyell, and others, it seems clearly established that earthquakes have their origin in some sudden impact of gas, steam, or molten matter impelled by gas or steam under high pressure, beneath the solid crust.[1] How such impact originates we need not stop to inquire, as the cause is closely connected with that which produces volcanic eruptions. The effect, however, of such impact is to originate a wave of translation through the crust, travelling outwards from the point, or focus, on the surface immediately over the point of impact.[2] These waves of translation can in some cases be laid down on a map, and are called "isoseismal curves," each curve representing approximately an equal degree of seismal intensity; as shown on the chart of a part of North America affected by the great Charleston earthquake. (Fig. 37.) Mr. Hopkins has shown that the earthquake-wave, when it passes through rocks differing in density and elasticity, changes in some degree not only its velocity, but its direction; being both refracted and reflected in a manner analogous to that of light when it passes from one medium to another of different density.[3] When a shock traverses the crust through a thickness of several miles it will meet with various kinds of rock as well as with fissures and plications of the strata, owing to which its course will be more or less modified.

(b.) Formation of Fissures.--During earthquake movements, fissures may be formed in the crust, and filled with gaseous or melted matter which may not in all cases reach the surface; and, on the principle that volcanoes are safety-valves for regions beyond their immediate influence, we may infer that the earthquake shock, which generally precedes the outburst of a volcano long dormant, finds relief by the eruption which follows; so that whatever may be the extent of the disastrous results of such an eruption, they would be still more disastrous if there had been no such safety-valve as that afforded by a volcanic vent. Thus, probably, owing to the extinction of volcanic activity in Syria, the earthquakes in that region have been peculiarly destructive. For example, on January 1, 1837, the town of Safed west of the Jordan valley was completely destroyed by an earthquake in which most of the inhabitants perished. The great earthquakes of Aleppo in the present century, and of Syria in the middle of the eighteenth, were of exceptional severity. In that of Syria, which took place in 1759, and which was protracted during a period of three months, an area of 10,000 square leagues was affected. Accon, Saphat, Baalbeck, Damascus, Sidon, Tripoli, and other places were almost entirely levelled to the ground; many thousands of human beings lost their lives.[4] Other examples might be cited.

(c.) Earthquake Waves.--We have now to return to the phenomena connected with the transmission of earthquake-waves. The velocity of transmission through the earth is very great, and several attempts have been made to measure this velocity with accuracy. The most valuable of such attempts are those connected with the Charleston and Riviera shocks. Fortunately, owing to the perfection of modern appliances, and the number of observers all over the globe, these results are entitled to great confidence. The phenomena connected with the Charleston earthquake, which took place on the 31st of August, 1886, are described in great detail by Captain Clarence E. Dutton, of the U.S. Ordnance Corps.[5] The conclusions arrived at are;--that as regards the depth of the focal point, this is estimated at twelve miles, with a probable error of less than two miles; while, as regards the rate of travel of the earthquake-wave, the estimate is (in one case) about 3.236 miles per second; and in another about 3.226 miles per second.

On the other hand, in the case of the earthquake of the Riviera, which took place on the 23rd of February, 1887, at 5.30 a.m. (local time), the vibrations of which appear to have extended across the Atlantic, and to have sensibly affected the seismograph in the Government Signal Office at Washington, the rate of travel was calculated at about 500 miles per hour, less than one-half that determined in the case of Charleston; but Captain Dutton claims, and probably with justice, that the results obtained in the latter case are far more reliable than any hitherto arrived at.

(d.) Oceanic Waves.--When the originating impact takes place under the bed of the ocean--either by a sudden up-thrust of the crust to the extent, let us suppose, of two or three feet, or by an explosion from a submarine volcano-- a double wave is formed, one travelling through the crust, the other through the ocean; and as the rate of velocity of the former is greatly in excess of that of the latter, the results on their reaching the land are often disastrous in the extreme. It is the ocean-wave, however, which is the more important, and calls for special consideration. If the impact takes place in very deep water, the whole mass of the water is raised in the form of a low dome, sloping equally away in all directions; and it commences to travel outwards as a wave with an advancing crest until it approaches the coast and enters shallow water. The wave then increases in height, and the water in front is drawn in and relatively lowered; so that on reaching a coast with a shelving shore the form of the surface consists of a trough in front followed by an advancing crest. These effects may be observed on a small scale in the case of a steamship advancing up a river, or into a harbour with a narrow channel, but are inappreciable in deep water, or along a precipitous open coast.

(e.) The Earthquake of Lisbon, 1755.--The disastrous results of a submarine earthquake upon the coast have never been more terribly illustrated than in the case of the earthquake of Lisbon which took place on November 1, 1755. The inhabitants had no warning of the coming danger, when a sound like that of thunder was heard underground, and immediately afterwards a violent shock threw down the greater part of their city; this was the land-wave. In the course of about six minutes, sixty thousand persons perished. The sea first retired and left the harbour dry, so forming the trough in front of the crest; immediately after the water rolled in with a lofty crest, some 50 feet above the ordinary level, flooding the harbour and portions of the city bordering the shore. The mountains of Arrabida, Estrella, Julio, Marvan, and Cintra, were impetuously shaken, as it were, from their very foundations; and according to the computation of Humboldt, a portion of the earth's surface four times the extent of Europe felt the effects of this great seismic shock, which extended to the Alps, the shores of the Baltic, the lakes of Scotland, the great lakes of North America, and the West Indian Islands. The velocity of the sea-wave was estimated at about 20 miles per minute.

(f.) Earthquake of Lima and Callao, 28th October, 1746.--Of somewhat similar character was the terrible catastrophe with which the cities of Lima and Callao were visited in the middle of the last century,[6] in which the former city, then one of great magnificence, was overthrown; and Callao was inundated by a sea-wave, in which out of 23 ships of all sizes in the harbour the greater number foundered; several, including a man-of-war, were lifted bodily and stranded, and all the inhabitants with the exception of about two hundred were drowned. A volcano in Lucanas burst forth the same night, and such quantities of water descended from the cone that the whole country was overflowed; and in the mountain near Pataz, called Conversiones de Caxamarquilla, three other volcanoes burst forth, and torrents of water swept down their sides. In the case of these cities, the land-wave, or shock, preceded the sea-wave, which of course only reached the port of Callao.

(g.) Earthquake of Charleston, 31st August, 1886.--I shall close this account of some remarkable earthquakes with a few facts regarding that of Charleston, on the Atlantic seaboard of Carolina.[7] At 9.51 a.m. of this day, the inhabitants engaged in their ordinary occupations were startled by the sound of a distant roar, which speedily deepened in volume so as to resemble the noise of cannon rattling along the road, "spreading into an awful noise, that seemed to pervade at once the troubled earth below and the still air above." At the same time the floors began to heave underfoot, the walls visibly swayed to and fro, and the crash of falling masonry was heard on all sides, while universal terror took possession of the populace, who rushed into the streets, the black portion of the community being the most demonstrative of their terror. Such was the commencement of the earthquake, by which nearly all the houses of Charleston were damaged or destroyed, many of the public buildings seriously injured or partially demolished. The effects were felt all over the States as far as the great lakes of Canada and the borders of the Rocky Mountains. Two epicentral foci appear to have been established; one lying about 15 miles to the N.W. of Charleston, called the Woodstock focus; the other about 14 miles due west of Charleston, called the Rantowles focus; around each of these foci the isoseismic curves concentrated,[8] but in the map (Fig. 37) are combined into the area of one curve. The position of these foci clearly shows that the origin of the Charleston earthquake was not submarine, though occurring within a short distance of the Atlantic border; the curves of equal intensity (isoseismals) are drawn all over the area influenced by the shock.

As a general result of these detailed observations, Captain Dutton states that there is a remarkable coincidence in the phenomena with those indicated by the theory of wave-motion as the proper one for an elastic, nearly homogeneous, solid medium, composed of such materials as we know to constitute the rocks of the outer portions of the earth; but on the other hand he states that nothing has been disclosed which seems to bring us any nearer to the precise nature of the forces which generated the disturbance.[9]

[1] The views of Mr. R. Mallet, briefly stated, are somewhat as follows:-- Owing to the secular cooling of the earth, and the consequent lateral crushing of the surface, this crushing from time to time overcomes the resistance; in which case shocks are experienced along the lines of fracture and faulting by which the crust is intersected. These shocks give rise to earthquake waves, and as the crushing of the walls of the fissure developes heat, we have here the vera causa both of volcanic eruptions and earthquake shocks--the former intensified into explosions by access of water through the fissures.--"On the Dynamics of Earthquakes," Trans. Roy. Irish Acad., vol. xxi.

[2] Illustration of the mode of propagation of earthquake shocks will be found in Lyell's Principles of Geology, vol. ii. p. 136, or in the author's Physiography, p. 76, after Hopkins.

[3] "Rep. on Theories of Elevation and Earthquakes," Brit. Ass. Rep. 1847, p. 33. In the map prepared by Prof. J. Milne and Mr. W. K. Burton to show the range of the great earthquake of Japan (1891), similar isoseismal lines are laid down.

[4] Lyell, loc. cit., p. 163. Two Catalogues of Earthquakes have been drawn up by Prof. O'Reilly, and are published in the Trans. Roy. Irish Academy, vol. xxviii. (1884 and 1886).

[5] Ninth Annual Report, U.S. Geological Survey (1888).

[6] A True and Particular Account of the Dreadful Earthquake, 2nd edit. The original published at Lima by command of the Viceroy. London, 1748. Translated from the Spanish.

[7] I take the account from that of Capt. Dutton above cited, p. 220.

[8] Dutton, Report, Plate xxvi., p. 308.

[9] Ibid., p. 211. On the connection between the moon's position and earthquake shocks, see Mr. Richardson's paper on Scottish earthquakes, Trans. Edin. Geol. Soc., vol. vi. p. 194 (1892).

PART VII.

VOLCANIC AND SEISMIC PROBLEMS.

CHAPTER I.

THE ULTIMATE CAUSE OF VOLCANIC ACTION.

Volcanic phenomena are the outward manifestations of forces deep-seated beneath the crust of the globe; and in seeking for the causes of such phenomena we must be guided by observation of their nature and mode of action. The universality of these phenomena all over the surface of our globe, in past or present times, indicates the existence of a general cause beneath the crust. It is true that there are to be found large tracts from which volcanic rocks (except those of great geological antiquity) are absent, such as Central Russia, the Nubian Desert, and the Central States of North America; but such absence by no means implies the non-existence of the forces which give rise to volcanic action beneath those regions, but only that the forces have not been sufficiently powerful to overcome the resistance offered by the crust over those particular tracts. On the other hand, the similarity of volcanic lavas over wide regions is strong evidence that they are drawn from one continuous magma, consisting of molten matter beneath the solid exterior crust.

(a.) Lines of Volcanic Action.--It has been shown in a previous page that volcanic action of recent or Tertiary times has taken place mainly along certain lines which may be traced on the surface of a map or globe. One of these lines girdles the whole globe, while others lie in certain directions more or less coincident with lines of flexure, plication or faulting. The Isle of Sumatra offers a remarkable example of the coincidence of such lines with those of volcanic vents. Not only the great volcanic cones, but also the smaller ones, are disposed in chains which run parallel to the longitudinal axis of the island (N.W.-S.E.). The sedimentary rocks are bent and faulted in lines parallel to the main axis, and also to the chains of volcanic mountains, and the observation holds good with regard to different geological periods.[1] Another remarkable case is that of the Jordan Valley. Nowhere can the existence of a great fracture and vertical displacement of the strata be more clearly determined than along this remarkable line of depression; and it is one which is also coincident with a zone of earthquake and volcanic disturbances.

(b.) Such Lines generally lie along the Borders of the Ocean.--But even where, from some special cause, actual observation on the relations of the strata are precluded, the general configuration of the ground and the relations of the boundaries between land and sea to those of volcanic chains, evidently point in many cases to their mutual interdependence. The remarkable straightness of the coast of Western America, and of the parallel chain of the Andes, affords presumptive evidence that this line is coincident with a fracture or system of faults, along which the continent has been bodily raised out of the waters of the ocean. Of this elevation within very recent times we have abundant evidence in the existence of raised coral-reefs and oceanic shell-beds at intervals all along the coast; rising in Peru to a level of no less than 3,000 feet above the ocean, as shown by Alexander Agassiz.[2] Such elevations probably occurred at a time when the volcanoes of the Andes were much more active than at present. Considered as a whole, these great volcanic mountains may be regarded as in a dormant, or partially moribund, condition; and if the volcanic forces have to some extent lost their strength, so it would appear have those of elevation.

(c.) Areas of Volcanic Action in the British Isles.--In the case of the British Islands it may be observed that the later Tertiary volcanic districts lie along very ancient depressions, which may indicate zones of weakness in the crust. Thus the Antrim plateau, as originally constituted, lay in the lap of a range of hills formed of crystalline, or Lower Silurian, rocks; while the volcanic isles of the Inner Hebrides were enclosed between the solid range of the Archan rocks of the Outer Hebrides on the one side, and the Silurian and Archan ranges of the mainland on the other. And if we go back to the Carboniferous period, we find that the volcanic district of the centre of Scotland was bounded by ranges of solid strata both to the north and south, where the resistance to interior pressure from molten matter would have been greater than in the Carboniferous hollow-ground, where such molten matter has been abundantly extruded. In all these cases, the outflow of molten matter was in a direction somewhat parallel to the plications of the strata.

(d.) Special Conditions under which the Volcanic Action operates.--Assuming, then, that the molten matter, forming an interior magma or shell, is constantly exerting pressure against the inner surface of the solid crust, and can only escape where the crust is too weak (owing to faults, plications, or fissures) to resist the pressure, we have to inquire what are the special conditions under which outbursts of volcanic matter take place, and what are the general results as regards the nature of the ejecta dependent on those conditions.

(e.) Effect of the Presence or Absence of Water.--The two chief conditions determining the nature of volcanic products, considered in the mass, are the presence or absence of water. Such presence or absence does not of course affect the essential chemical composition of the ejecta, but it materially influences the form in which the matter is erupted. The agency of water in volcanic eruptions is a very interesting and important subject in connection with the history of volcanic action, and has been ably treated by Professor Prestwich.[3] At one time it was considered that water was essential to volcanic activity; and the fact that the great majority of volcanic cones are situated in the vicinity of the oceanic waters, or of inland seas, was pointed to in confirmation of this theory. But the existence in Western America and other volcanic countries of fissures of eruption along which molten lava has been extruded without explosions of steam, shows that water is not an essential factor in the production of volcanic phenomena; and, as Professor Prestwich has clearly demonstrated, it is to be regarded as an element in volcanic explosions, rather than as a prime cause of volcanic action. The main difficulty he shows to be thermo-dynamical; and calculating the rate of increase in the elastic force of steam on descending to greater and greater depths and reaching strata of higher and higher temperatures, as compared with the force of capillarity, he comes to the conclusion that water cannot penetrate to depths of more than seven or eight miles, and therefore cannot reach the seat of the eruptive forces. Professor Prestwich also points out that if the extrusion of lava were due to the elastic force of vapour of water there should be a distinct relation between the discharge of the lava and of the vapour; whereas the result of an examination of a number of well-recorded eruptions shows that the two operations are not related, and are, in fact, perfectly independent. Sometimes there has been a large discharge of lava, and little or no escape of steam; at other times there have been paroxysmal explosive eruptions with little discharge of lava. Even in the case of Vesuvius, which is close to the sea, there have been instances when the lava has welled out almost with the tranquillity of a water-spring.

(f.) Access of Surface Water to Molten Lava during Eruptions.--The existence of water during certain stages in eruptions is too frequent a phenomena to be lost sight of; but its presence may be accounted for in other ways, besides proximity to the sea or ocean. Certain volcanic mountains, such as Etna and Vesuvius, are built upon water-bearing strata, receiving their supplies from the rainfall of the surrounding country, or perhaps partly from the sea. In addition to this the ashes and scori?of the mountain sides are highly porous, and rain or snow can penetrate and settle downwards around the pipe or throat through which molten lava wells up from beneath. In such cases it is easy to understand how, at the commencement of a period of activity, molten lava ascending through one or more fissures, and meeting with water-charged strata or scori? will convert the water into steam at high pressure, resulting in explosions more or less violent and prolonged, in proportion to the quantity of water and the depth to which it has penetrated. In this manner we may suppose that ashes, scori? and blocks of rock torn from the sides of the crater-throat, and hurled into the air, are piled around the vent, and accumulate into hills or mountains of conical form. After the explosion has exhausted itself, the molten lava quietly wells up and fills the crater, as in the cases of those of Auvergne and Syria, and other places. We may, therefore, adopt the general principle that in volcanic eruptions where water in large quantities is present, we shall have crater-cones built up of ashes, scori? and pumice; but where absent, the lava will be extravasated in sheets to greater or less distances without the formation of such cones; or if cones are fanned, they will be composed of solidified lava only, easily distinguishable from crater-cones of the first class.

(g.) Nature of the Interior Reservoir from which Lavas are derived.--We have now to consider the nature of the interior reservoir from which lavas are derived, and the physical conditions necessary for their eruption at the surface.

Without going back to the question of the original condition of our globe, we may safely hold the view that at a very early period of geological history it consisted of a solidified crust at a high temperature, enfolding a globe of molten matter at a still higher temperature. As time went on, and the heat radiated into space from the surface of the globe, while at the same time slowly ascending from the interior by conduction, the crust necessarily contracted, and pressing more and more on the interior molten magma, this latter was forced from time to time to break through the contracting crust along zones of weakness or fissures.

(h.) The Earth's Crust in a State of both Exterior Thrust and of Interior Tension.--As has been shown by Hopkins,[4] and more recently by Mr. Davison,[5] an exterior crust in such a condition must eventually result in being under a state of horizontal thrust towards the exterior and of tension towards the interior surface. For the exterior portion, having cooled down, and consequently contracted to its normal state, will remain rigid up to a certain point of resistance; but the interior portion still continuing to contract, owing to the conduction of the heat towards the exterior, would tend to enter upon a condition of tension, as becoming too small for the interior molten magma; and such a state of tension would tend to produce rupture of the interior part. In this manner fissures would be formed into which the molten matter would enter; and if the fissures happened to extend to the surface, owing to weakness of the crust or flexuring of the strata, or other cause, the molten matter would be extruded either in the form of dykes or volcanic vents. In this way we may account for the numerous dykes of trap by which some volcanic districts are intersected, such as those of the north of Ireland and centre of Scotland.

From the above considerations, it follows that the earth's crust must be in a condition both of pressure (or lateral thrust) towards the exterior portion, and of tension towards the interior, the former condition resulting in faulting and flexuring of the rocks, the latter in the formation of open fissures, through which lava can ascend under high pressure. These operations are of course the attempt of the natural forces to arrive at a condition of equilibrium, which is never attained because the processes are never completed; in other words, radiation and convection of heat are constantly proceeding, giving rise to new forces of thrust and tension.

It now remains for us to consider what may be the condition of the interior molten magma; and in doing so we must be guided to a large extent by considerations regarding the nature of the extruded matter at the surface.

(i.) Relative Densities of Lavas.--Now, observation shows that, as bearing on the subject under consideration, lavas occur mainly under two classes as regards their density. The most dense (or basic) are those in which silica is deficient, but iron is abundant; the least dense (or acid) are those which are rich in silica, but in which iron occurs in small quantity. This division corresponds with that proposed by Bunsen and Durocher[6] for volcanic rocks, upon the results of analyses of a large number of specimens from various districts. Rocks may be thus arranged in groups:

(1) The Basic (Heavier)--poor in silica, rich in iron; containing silica 45-58 per cent. Examples: Basalt, Dolerite, Hornblende rock, Diorite, Diabase, Gabbro, Melaphyre, and Leucite lava.

(2) The Acid (Lighter)--rich in silica, poor in iron; containing silica 62-78 per cent. Examples: Trachyte, Rhyolite, Obsidian, Domite, Felsite, Quartz-porphyry, Granite.

The Andesite group forms a connecting link between the highly acid and the basic groups, and there are many varieties of the above which it is not necessary to enumerate. Durocher supposes that the molten magmas of these various rocks are arranged in concentric shells within the solid crust in order of their respective densities, those of the lighter density, namely the acid magmas, being outside those of greater density, namely the basic; and this is a view which seems not improbable from a consideration not only of the principle itself, but of the succession of the varieties of lava in many districts. Thus we find that acid lavas have been generally extruded first, and basic afterwards--as in the cases of Western America, of Antrim, the Rhine and Central France. And if the interior of our globe had been in a condition of equilibrium from the time of the consolidation of the crust to the present, reason would induce us to conclude that the lavas would ultimately have arranged themselves in accordance with the conditions of density beneath that crust. But the state of equilibrium has been constantly disturbed. Every fresh outburst of volcanic force, and every fresh extrusion of lava, tends to disturb it, and to alter the relations of the interior viscous or molten magmas. Owing to this it happens, as we may suppose, that the order of eruption according to density is sometimes broken, and we find such rocks as granophyre (a variety of andesite) breaking through the plateau-basalts of Mull and Skye, as explained in a former chapter. Notwithstanding such variations, however, the view of Durocher may be considered as the most reasonable we can arrive at on a subject which is confessedly highly conjectural.

(j.) Conclusion as regards the Ultimate Cause of Volcanic Action.-- Notwithstanding, however, the complexity of the subject, and the uncertainties which must attend an inquiry where some of the data are outside the range of our observation, sufficient evidence can be adduced to enable us to arrive at a tolerably clear view of the ultimate cause of volcanic action. So tempting a subject was sure to evoke numerous essays, some of great ingenuity, such as that of Mr. Mallet; others of great complexity, such as that of Dr. Daubeny. But more recent consideration and wider observation have tended to lead us to the conclusion that the ultimate cause is the most simple, the most powerful, and the most general which can be suggested; namely, the contraction of the crust due to secular cooling of the more deeply seated parts by conduction and radiation of heat into space. Owing to this cause, the enclosed molten matter is more or less abundantly extruded from time to time along the lines and vents of eruption, so as to accommodate itself to the ever-contracting crust. Nor can we doubt that this process has been going on from the very earliest period of the earth's history, and formerly at a greater rate than at present. When the crust was more highly heated, the radiation and conduction must have been proportionately more rapid. Owing to this cause also the contraction of the crust was accelerated. To such irresistible force we owe the wonderful flexuring, folding, and horizontal overthrusting which the rocks have undergone in some portions of the globe--such as in the Alps, the Highlands of Scotland and of Ireland, and the Alleghannies of America. It is easy to show that the acceleration of the earth's rotation must be a consequence of such contraction; but, after all, this is but one of those compensatory forces of which we see several examples in the world around us. It can also be confidently inferred that at an early period of the earth's history, when the moon was nearer to our planet than at present, the tides were far more powerful, and their effect in retarding the earth's rotation was consequently greater. During this period the acceleration due to contraction was also greater; and the two forces probably very nearly balanced each other. Both these forces (those of acceleration and retardation) have been growing weaker down to the present day, though there appears to have been a slight advantage on the side of the retarding force.[7]

[1] R. D. M. Verbeek, Krakatau, p. 105 (1886); also, J. Milne, The Great Earthquake of Japan, 1891.

[2] Bull. Mus. Comp. Zool., vol. iii.

[3] Proc. Roy. Soc., No. 237 (1885); also, Rep. Brit. Assoc. (1881).

[4] Hopkins, supra cit., p. 218.

[5] C. Davison and G. H. Darwin, Phil. Trans., vol. 178, p; 241.

[6] Durocher, Ann. des Mines, vol. ii. (1857).

[7] See on this subject the author's Textbook of Physiography (Deacon and Co., 1888), pp. 56 and 122.

CHAPTER II.

LUNAR VOLCANOES.

The surface of the moon presented to our view affords such remarkable indications of volcanic phenomena of a special kind, that we are justified in devoting a chapter to their consideration. It is very tantalising that our beautiful satellite only permits us to look at and admire one half of her sphere; but it is not a very far-fetched inference if we feel satisfied that the other half bears a general resemblance to that which is presented to the earth. It is scarcely necessary to inform the reader why it is that we never see but one face; still, for the sake of those who have not thought out the subject I may state that it is because the moon rotates on her axis exactly in the time that she performs a revolution round the earth. If this should not be sufficiently clear, let the reader perform a very simple experiment for himself, which will probably bring conviction to his mind that the explanation here given is correct. Let him place an orange in the centre of a round table, and then let him move round the table from a starting-point sideways, ever keeping his face directed towards the orange; and when he has reached his starting-point, he will find that he has rotated once round while he has performed one revolution round the table. In this case the performer represents the moon and the orange the earth.

Now this connection between the earth and her satellite is sufficiently close to be used as an argument (if not as actual demonstration) that the earth and the moon were originally portions of the same mass, and that during some very early stage in the development of the solar system these bodies parted company, to assume for ever after the relations of planet and satellite. At the epoch referred to, we may also suppose that these two masses of matter were in a highly incandescent, if not even gaseous, state; and we conclude, therefore, that having been once portions of the same mass, they are composed of similar materials. This conclusion is of great importance in enabling us to reason from analogy regarding the origin of the physical features on the moon's surface, and for the purpose of comparison with those which we find on the surface of our globe; because it is evident that, if the composition of the moon were essentially different from that of our earth, we should have no basis whatever for a comparison of their physical features.

When the moon started on her career of revolution round the earth, we may well suppose that her orbit was much smaller than at present. She was influenced by counteracting forces, those of gravitation drawing her towards the centre of gravity of the earth,[1] and the centrifugal force, which in the first instance was the stronger, so that her orbit for a lengthened period gradually increased until the two forces, those of attraction and repulsion, came into a condition of equilibrium, and she now performs her revolution round the earth at a mean distance of 240,000 miles, in an orbit which is only very slightly elliptical.[2] How the period of the moon's rotation is regulated by the earth's attraction on her molten lava-sheets, first at the surface, and now probably below the outer crust, has been graphically shown by Sir Robert Ball,[3] but it cannot be doubted that once the moon was appreciably nearer to our globe than at present. The attraction of her mass produced tides in the ocean of correspondingly greater magnitude, and capable of effecting results, both in eroding the surface and in transporting masses of rock, far beyond the bounds of our every-day experience.

Of all the heavenly bodies, the sun excepted, the moon is the most impressive and beautiful. As we catch her form, rising as a fair crescent in the western sky after sunset, gradually increasing in size and brilliancy night after night till from her circular disk she throws a full flood of light on our world and then passes through her decreasing phases, we recognise her as "the Governor of the night," or in the words of our own poet, when in her crescent phase, "the Diadem of night." Seen through a good binocular glass, her form gains in rotundity; but under an ordinary telescope with a four-inch objective, she appears like a globe of molten gold. Yet all this light is derivative, and is only a small portion of that she receives from the sun. That her surface is a mass of rigid matter destitute of any inherent brilliancy, appears plain enough when we view a portion of her disk through a very large telescope. It was the good fortune of the author to have an opportunity for such a view through one of the largest telescopes in the world. The 27-inch refractor manufactured by Sir Howard Grubb of Dublin, for the Vienna observatory, a few years ago, was turned on a portion of the moon's disk before being finally sent off to its destination; and seen by the aid of such enormous magnifying power, nothing could be more disappointing as regards the appearance of our satellite. The sheen and lustre of the surface was now observed no longer; the mountains and valleys, the circular ridges and hollows were, indeed, wonderfully defined and magnified, but the matter of which they seemed to be constituted resembled nothing so much as the pale plaster of a model. One could thus fully realise the fact that the moon's light is only derivative. Still we must recollect that the most powerful telescope can only bring the surface of the moon to a distance from us of about 250 miles; and it need not be said that objects seen at such a distance on our earth present very deceptive appearances; so that we gain little information regarding the composition of the moon's crust, or exterior surface, simply from observation by the aid of large telescopes.

Reasoning from analogy with our globe, we may infer that the exterior shell of the moon consists of crystalline volcanic matter of the highly silicated, or acid, varieties resting upon another of a denser description, rich in iron, and resembling basalt. This hypothesis is hazarded on the supposition that the composition of the matter of the moon's mass resembles in the main that of our globe. During the process of cooling from a molten condition, the heavier lavas would tend to fall inwards, and allow the lighter to come to the surface, and form the outer shell in both cases. Thus, the outer crust would resemble the trachytic lavas of our globe, and their pale colour would enable the sun's rays to be reflected to a greater extent than if the material were of the blackness of basalt.[4] So much for the material. We have now to consider the structure of the moon's surface, and here we find ourselves treading on less speculative and safer ground. All astronomers since the time of Schroter seem to be of accord in the opinion that the remarkable features of the moon's surface are in some measure of volcanic origin, and we shall presently proceed to consider the character of these forms more in detail.

But first, and as leading up to the discussion of these physical features, we must notice one essential difference between the constitution of the moon and of the earth; namely, the absence of water and of an atmosphere in the case of the moon. The sudden and complete occultation of the stars when the moon's disk passes between them and the place of the observer on the earth's surface, is sufficient evidence of the absence of air; and, as no cloud has ever been noticed to veil even for a moment any part of our satellite's face, we are pretty safe in concluding that there is no water; or at least, if there be any, that it is inappreciable in quantity.[5] Hence we infer that there is no animal or vegetable life on the moon's surface; neither are there oceans, lakes or rivers, snowfields or glaciers, river-valleys or canyons, islands, stratified rocks, nor volcanoes of the kind most prevalent on our own globe.

[Illustration: Fig. 38.--Photograph of the moon's surface (in part) showing the illuminated "spots," and ridges, and the deep hollows. The position of "Tycho" is shown near the upper edge, and some of the volcanic craters are very clearly seen near the margin.]

Now on looking at a photographic picture of the moon's surface (Fig. 38), we observe that there are enormous dark spaces, irregular in outline, but more or less approaching the circular form, surrounded by steep and precipitous declivities, but with sides sloping outwards. These were supposed at one time to be seas; and they retain the name, though it is universally admitted that they contain no water. Some of these hollows are four English miles in depth. The largest of these, situated near the north pole of the moon, is called Mare Imbrium; next to it is Mare Serenitatis; next, Mare Tranquilitatis, with several others.[6] Mare Imbrium is of great depth, and from its floor rise several conical mountains with circular craters, the largest of which, Archimedes, is fifty miles in diameter; its vast smooth interior being divided into seven distinct zones running east and west. There is no central mountain or other obvious internal sign of former volcanic activity, but its irregular wall rises into abrupt towers, and is marked outside by decided terraces.[7]

The Mare Imbrium is bounded along the east by a range of mountains called the Apennines, and towards the north by another range called the Alps; while a third range, that of the Caucasus, strikes northward from the junction of the two former ranges. Several circular or oval craters are situated on, and near to, the crest of these ridges.

[Illustration: Fig. 39.--A magnified portion of the moon's surface, showing the forms of the great craters with their outer ramparts. The white spot with shadow is a cone rising from the centre of one of the larger craters to a great height and thus becoming illuminated by the sun's light.]

But the greater part of the moon's hemisphere is dotted over by almost innumerable circular crater-like hollows; sometimes conspicuously surmounting lofty conical mountains, at other times only sinking below the general outer surface of the lunar sphere. On approaching the margin, these circular hollows appear oval in shape owing to their position on the sphere; and the general aspect of those that are visible leads to the conclusion that there are large numbers of smaller craters too small to be seen by the most powerful telescopes. These cones and craters are the most characteristic objects on the whole of the visible surface, and when highly magnified present very rugged outlines, suggestive of slag, or lava, which has consolidated on cooling, as in the case of most solidified lava-streams on our earth.[8] One of the most remarkable of these crateriform mountains is that named Copernicus, situated in a line with the southern prolongation of the Apennines. Of this mountain Sir R. Ball says: "It is particularly well known through Sir John Herschel's drawing, so beautifully reproduced in the many editions of the Outlines of Astronomy. The region to the west is dotted over with innumerable minute craterlets. It has a central, many-peaked mountain about 2,400 feet in height. There is good reason to believe that the terracing shown in its interior is mainly due to the repeated alternate rise, partial congealation and retreat of a vast sea of lava. At full moon it is surrounded by radiating streaks."[9] The view regarding the structure of Copernicus here expressed is of importance, as it is probably applicable to all the craters of our satellite.

"When the moon is five or six days old," says Sir Robert Ball, "a beautiful group of three craters will be readily found on the boundary line between night and day. These are Catharina, Cyrillus, and Theophilus. Catharina is the most southerly of the group, and is more than 16,000 feet deep and connected to Cyrillus by a wide valley; but between Cyrillus and Theophilus there is no such connection. Indeed Cyrillus looks as if its huge surrounding ramparts, as high as Mont Blanc, had been completely finished when the volcanic forces commenced the formation of Theophilus, the rampart of which encroaches considerably on its older neighbour. Theophilus stands as a well-defined round crater, about 64 miles in diameter, with an internal depth of 14,000 to 18,000 feet, and a beautiful central group of mountains, one-third of that height, on its floor. This proves that the last eruptive efforts in this part of the moon fully equalled in intensity those that had preceded them. Although Theophilus is on the whole the deepest crater we can see in the moon, it has received little or no deformation by secondary eruptions."

But perhaps the most remarkable object on the whole hemisphere of the moon is "the majestic Tycho," which rises from the surface near the south pole, and at a distance of about 1/6th of the diameter of the sphere from its margin. Its depth is stated by Ball to be 17,000 feet, and its diameter 50 miles. But its special distinction amongst the other volcanic craters lies in the streaks of light which radiate from it in all directions for hundreds and even thousands of miles, stretching with superb indifference across vast plains, into the deepest craters, and over the highest opposing ridges. When the sun rises on Tycho these streaks are invisible, but as soon as it has reached a height of 25?to 30?above the horizon, the rays emerge from their obscurity, and gradually increase in brightness until full moon, when they become the most conspicuous objects on her surface. As yet no satisfactory explanation has been given of the origin of these illuminated rays,[10] but I may be permitted to add that their form and mode of occurrence are eminently suggestive of gaseous exhalations from the volcano illumined by the sun's rays; and owing to the absence of an atmosphere, spreading themselves out in all directions and becoming more and more attenuated until they cease to be visible.

The above account will probably suffice to give the reader a general idea of the features and inferential structure of the moon's surface. That she was once a molten mass is inferred from her globular form; but, according to G. F. Chambers, the most delicate measurements indicate no compression at the poles.[11] That her surface has cooled and become rigid is also a necessary inference; though Sir J. Herschel considered that the surface still retains a temperature possibly exceeding that of boiling water.[12] However this may be, it is pretty certain that whatever changes may occur upon her surface are not due to present volcanic action, all evidence of such action being admittedly absent. If, when the earth and moon parted company, their respective temperatures were equal, the moon being so much the smaller of the two would have cooled more rapidly, and the surface may have been covered by a rigid crust when as yet that of the earth may have been molten from heat. Hence the rigidity of the moon's surface may date back to an immensely distant period, but she may still retain a high temperature within this crust. Having arrived at this stage of our narrative, we are in a position to consider by what means, and under what conditions, the cones and craters which diversify the lunar surface have been developed.

In doing so it may be desirable, in the first place, to determine what form of crater on our earth's surface those of the moon do not represent; and we are guided in our inquiry by the consideration of the absence of water on the lunar surface. Now there are large numbers of crateriform mountains on our globe in the formation of which water has played an important, indeed essential, part. As we have already seen, water, though not the ultimate cause of volcanic eruptions, has been the chief agent, when in the form of steam at high pressure, in producing the explosions which accompany these

eruptions, and in tearing up and hurling into the air the masses of rock, scori?
and ashes, which are piled around the vents of eruption in the form of craters
during periods of activity. To this class of craters those of Etna, Vesuvius, and
Auvergne belong. These mountains and conical hills (the domes excepted) are
all built up of accumulations of fragmental material, with occasional sheets
and dykes of lava intervening; and where eruptions have taken place in
recent times, observation has shown that they are accompanied by outbursts
of vast quantities of aqueous vapour, which has been the chief agent (along
with various gases) in piling up the circular walls of the crater.

It has also been shown that in many instances these crater-walls have been
breached on one side, and that streams of molten lava which once occupied
the cup to a greater or less height, have poured down the mountain side.
Hence the form or outline of many of these fragmental craters is crescent-
shaped. Such breached craters are to be found in all parts of the world, and
are not confined to any one district, or even continent, so that they may be
considered as characteristic of the class of volcanic crater-cones to which I am
now referring. In the case of the moon, however, we fail to observe any
decided instances of breached craters, with lava-streams, such as those I
have described.[13] In nearly all cases the ramparts appear to extend
continuously round the enclosed depression, solid and unbroken; or at least
with no large gap occupying a very considerable section of the circumference.
(See Fig. 38.) Hence we are led to suspect that there is some essential
distinction between the craters on the surface of the moon and the greater
number of those on the surface of our earth.

It is scarcely necessary to add that the volcanic mountains of the moon offer
no resemblance whatever to the dome-shaped volcanic mountains of our
globe. If it were otherwise, the lunar mountains would appear as simple
luminous points rising from a dark floor, over which they would cast a conical
shadow. But the form of the lunar volcanic mountains is essentially different;
as already observed, they consist in general of a circular rampart enclosing a
depressed floor, sometimes terraced as in the case of Copernicus, from which
rise one or more conical mountains, which are in effect the later vents of
eruption.

In our search, therefore, for analogous forms on our own earth, we must
leave out the craters and domes of the type furnished by the European

volcanoes and their representatives abroad, and have recourse to others of a different type. Is there then, we may ask, any type of volcanic mountain on our globe comparable with those on the moon? In all probability there is.

If the reader will turn to the description of the volcanoes of the Hawaiian group in the Pacific, especially that of Mauna Loa, as given by Professor Dana and others, and compare it with that of Copernicus, he will find that in both cases we have a circular rampart of solid lava enclosing a vast plain of the same material from which rise one or more lava-cones. The interiors in both cases are terraced. So that, allowing for differences in magnitude, it would seem that there is no essential distinction between lunar craters and terrestrial craters of the type of Mauna Loa. Dana calls these Hawaiian volcanoes "basaltic," basalt being the prevalent material of which they are formed. Those of the moon may be composed of similar material, or otherwise; but in either case we may suppose they are built up of lava, erupted from vents connected with the molten reservoirs of the interior. Thus we conclude that they belong to an entirely different type, and have been built up in a different manner, from those represented by Etna, Vesuvius, and most of the extinct volcanoes of Auvergne, the Eifel, and of other districts considered in these pages.

Let us now endeavour to picture to ourselves the stages through which the moon may be supposed to have passed from the time her surface began to consolidate owing to the radiation of her heat into space; for there is every probability that some of the craters now visible on her disk were formed at a very early period of her physical history.

When the surface began to consolidate, it must also have contracted; and the interior molten matter, pressed out by the contracting crust, must have been over and over again extruded through fissures produced over the solidified surface, until the solid crust extended over the whole lunar surface, and became of considerable thickness.

It is from this epoch that, in all probability, we should date the commencement of what may be termed "the volcanic history" of the moon. We must bear in mind that although the moon's surface had become solid, its temperature may have remained high for a very long period. But the continuous radiation of the surface-heat into space would produce continuous contraction, while the convection of the interior heat would tend to increase the thickness of the outer solid shell; and this, ever pressing with increasing force on the interior molten mass, would result in frequent ruptures of the shell, and the extrusion of molten lava rising from below. Hence we may suppose the fissure-eruptions of lava were of frequent occurrence for a lengthened period during the early stage of consolidation of the lunar crust; but afterwards these may be supposed to have given place to eruptions through pipes or vents, resulting in the formation of the circular craters which form such striking and characteristic objects in the physical aspect of our satellite.[14]

It is not to be supposed that the various physical features on the lunar surface have all originated in the same way. The great ranges of mountains previously described may have originated by a process of piling up of immense masses of molten lava extruded from the interior through vents or fissures; while the great hollows (or "seas") are probably due to the falling inwards of large spaces owing to the escape of the interior lava.

But it is with the circular craters that we are most concerned. Judging from analogy with the lava-craters present on our globe, we must suppose them to be due to the extrusion, and piling up, of lava through central pipes, followed in some cases by the subsidence of the floor of the crater. It seems not improbable that it was in this way the greater number of the circular craters lying around Tycho, and dotting so large a space round the margin of the moon, were constructed. (See Fig. 38.) In general they appear to consist of an elevated rim, enclosing a depressed plain, out of which a central cone arises. The rim may be supposed to have been piled up by successive discharges of lava from a central orifice; and after the subsidence of the paroxysm the lava still in a molten condition may have sunk down, forming a seething lake within the vast circular rampart, as in the case of the Hawaiian volcanoes. The terraces observable within the craters in some instances have probably been left by subsequent eruptions which have not attained to the level of preceding ones; and where a central crater-cone is seen to rise within the caldron, we may suppose this to have been built up by a later series of eruptions of lava through the original pipe after the consolidation of the interior sea of lava. The mamelons of the Isle of Bourbon,[15] and some of the lava-cones of Hawaii, appear to offer examples on our earth's surface of these peculiar forms.

Such are the views of the origin of the physical features of our satellite which their form and inferred constitution appear to suggest. They are not offered with any intention of dogmatising on a subject which is admittedly obscure, and regarding which we have by no means all the necessary data for coming to a clear conclusion. All that can be affirmed is, that there is a great deal to be said in support of them, and that they are to some extent in harmony with phenomena within range of observation on the surface of our earth.

The far greater effects of lunar vulcanicity, as compared with those of our globe, may be accounted for to some extent by the consideration that the force of gravity on the surface of the moon is only one-sixth of that on the surface of the earth. Hence the eruptive forces of the interior of our satellite have had less resistance to overcome than in the case of our planet; and the erupted materials have been shot forth to greater distances, and piled up in greater magnitude, than with us. We have also to recollect that the abrading action of water has been absent from the moon; so that, while accumulations of matter had been proceeding throughout a prolonged period over its surface, there was no counteracting agency of denudation at work to modify or lessen the effects of the ruptive forces.

[1] Correctly speaking, each attracts the other towards its centre of gravity with a force proportionate to its mass, and inversely as the square of the distance; but the earth being by much the larger body, its attraction is far greater than that of the moon.

[2] The variation in the distance is only under rare circumstances 40,000 miles, but ordinarily about 13,000 miles.

[3] Story of the Heavens, 2nd edition, p. 525, et seq.

[4] A series of researches made by Zulner, of Leipzig, led him to assign to the light-reflecting capacity of the full-moon a result intermediate between that obtained by Bouguer, which gave a brightness equal to 1/300000 part of that of the sun, and of Wollaston, which gave 1/801070 part. We may accept 1/618000 of Zulner as sufficiently close; so that it would require 600,000 full moons to give the same amount of light as that of the sun.

[5] Schroter, however, came to the conclusion that the moon has an atmosphere.

[6] A chart of the moon's surface, with the names of the principal physical features, will be found in Ball's Story of the Heavens, 2nd edit., p. 60. It must be remembered that the moon as seen through a telescope appears in reversed position.

[7] Ibid., p. 66.

[8] As represented by Nasmyth's models in plaster.

[9] Ball, loc. cit., p. 67.

[10] Ball, loc. cit., p. 69.

[11] Astronomy, p. 78.

[12] Outlines of Astronomy, p. 285.

[13] At rare intervals a few crescent-shaped ridges are discernible on the lunar sphere, but it is very doubtful if they are to be regarded as breached craters.

[14] The number of "spots" on the moon was considered to be 244 until Schroter increased it to 6,000, and accurately described many of them. Schroter seems to have been the earliest observer who identified the circular hollows on the moon's surface as volcanic craters.

[15] Drawings of these very curious forms are given by Judd, Volcanoes, p. 127.

CHAPTER III.

ARE WE LIVING IN AN EPOCH OF SPECIAL VOLCANIC ACTIVITY?

The question which we are about to discuss in the concluding chapter of this volume is one to which we ought to be able to offer a definite answer. This can only be arrived at by a comparison of the violence and extent of volcanic and seismic phenomena within the period of history with those of pre-historic periods.

At first sight we might be disposed to give to the question an affirmative reply when we remember the eruptions of the last few years, and add to these the volcanic outbursts and earthquake shocks which history records. The cases of the earthquake and eruption in Japan of November, 1891, where in one province alone two thousand people lost their lives and many thousand houses were levelled[1]; that of Krakatoa, in 1883; of Vesuvius, in 1872; and many others of recent date which might be named, added to those which history records;--the recollection of such cases might lead us to conclude that our epoch is one in which the subterranean volcanic forces had broken out with extraordinary energy over the earth's surface. Still, when we come to examine into the cases of recorded eruptions--especially those of great violence--we find that they are limited to very special districts; and even if we extend our retrospect into the later centuries of our era, we shall find that the exceptionally great eruptions have been confined to certain permanently volcanic regions, such as the chain of the Andes, that of the Aleutian, Kurile, Japanese, and Philippine and Sunda Islands, lying for the most part along the remarkable volcanic girdle of the world to which I have referred in a previous page. Add to these the cases of Iceland and the volcanic islands of the Pacific, and we have almost the whole of the very active volcanoes of the world.

Then for the purposes of our inquiry we have to ascertain how these active vents of eruption compare, as regards the magnitude of their operations, with those of the pre-historic and later Tertiary times. But before entering into this question it maybe observed, in the first place, that a large number of the vents of eruption, even along the chain of the earth's volcanic girdle, are dormant or extinct. This observation applies to many of the great cones and domes of the Andes, including Chimborazo and other colossal mountains in Ecuador, Columbia, Chili, Peru, and Mexico. The region between the eastern Rocky Mountains and the western coast of North America was, as we have seen, one over which volcanic eruptions took place on a vast scale in later Tertiary times; but one in which only the after-effects of volcanic action are at present in operation. We have also seen that the chain of volcanoes of Japan and of the Kurile Islands are only active to a slight extent as compared with former times, and the same observation applies to those of New Zealand. Out of 130 volcanoes in the Japanese islands, only 48 are now believed to be active.

Again, if we turn to other districts we have been considering, we find that in the Indian Peninsula, in Arabia, in Syria and the Holy Land, in Persia, in Abyssinia and Asia Minor--regions where volcanic operations were exhibited on a grand scale throughout the Tertiary period, and in some cases almost down into recent times--we are met by similar evidences either of decaying volcanic energy, or of an energy which, as far as surface phenomena are concerned, is a thing of the past. Lastly, turning our attention to the European area, notwithstanding the still active condition of Etna, Vesuvius, and a few adjoining islands, we see in all directions throughout Southern Italy evidences of volcanic operations of a past time,--such as extinct crater-cones, lakes occupying the craters of former volcanoes, and extensive deposits of tuff or streams of lava--all concurring in giving evidence of a period now past, when vulcanicity was widespread over regions where its presence is now never felt except when some earthquake shock, like that of the Riviera, brings home to our minds the fact that the motive force is still beneath our feet, though under restrained conditions as compared with a former period.

Similar conclusions are applicable with even greater force to other parts of the European area. The region of the Lower Rhine and Moselle, of Hungary and the Carpathians, of Central France, of the North of Ireland and the Inner Hebrides, all afford evidence of volcanic operations at a former period on an extensive scale; and the contrast between the present physically silent and peaceful condition of these regions, as regards any outward manifestations of sub-terrestrial forces, compared with those which were formerly prevalent, cannot fail to impress our minds irresistibly with the idea that volcanic energy has well-nigh exhausted itself over these tracts of the earth's surface.

From this general survey of the present condition of the earth's surface, as regards the volcanic operations going on over it, and a comparison with those of a preceding period, we are driven to the conclusion that, however violent and often disastrous are the volcanic and seismic phenomena of the present day, they are restricted to comparatively narrow limits; and that even within these limits the volcanic forces are less powerful than they were in prehistoric times.

The middle part of the Tertiary period appears, in fact, to have been one of extraordinary volcanic activity, whether we regard the wide area over which this activity manifested itself, or the results as shown by the great amount of the erupted materials. Many of the still active volcanic chains, or groups, probably had their first beginnings at the period referred to; but in the majority of cases the eruptive forces have become dormant or extinct. With the exception of the lavas of the Indian-Peninsular area, which appear, at least partially, to belong to the close of the Cretaceous epoch, the specially volcanic period may be considered to extend from the beginning of the Miocene down to the close of the Pliocene stage. During the Eocene stage, volcanic energy appears to have been to a great degree dormant; but plutonic energy was gathering strength for the great effort of the Miocene epoch, when the volcanic forces broke out with extraordinary violence over Europe, the British Isles, and other regions, and continued to develop throughout the succeeding Pliocene epoch, until the whole globe was surrounded by a girdle of fire.

* * * * *

The reply, therefore, to the question with which we set out is very plain; and is to the effect that the present epoch is one of comparatively low volcanic activity. The further question suggests itself, whether the volcanic phenomena of the middle Tertiary period bear any comparison with those of past geological times. This, though a question of great interest, is one which is far too large to be discussed here; and it is doubtful if we have materials available upon which to base a conclusion. But it may be stated with some confidence, in general terms, that the history of the earth appears to show that, throughout all geological time, our world has been the theatre of intermittent geological activity, periods of rest succeeding those of action; and if we are to draw a conclusion regarding the present and future, it would be that, owing to the lower rate of secular cooling of the crust, volcanic action ought to become less powerful as the world grows older.

[1] Admirably illustrated in Prof. J. Milne's recently published work, The Great Earthquake of Japan, 1891.

APPENDIX.

A BRIEF ACCOUNT OF THE PRINCIPAL VARIETIES OF VOLCANIC ROCKS.

The text-books on this subject are so numerous and accessible, that a very brief account of the volcanic rocks is all that need be given here for the purposes of reference by readers not familiar with petrological details.

Let it be observed, in the first place, that there is no hard and fast line between the varieties of igneous and volcanic rocks. In this as in other parts of creation--natura nil facit per saltum; there are gradations from one variety to the other. At the same time a systematic arrangement is not only desirable, but necessary; and the most important basis of arrangement is that founded on the proportion of silica (or quartz) in the various rocks, as first demonstrated by Durocher and Bunsen, who showed that silica plays the same part in the inorganic kingdom that carbon does in the organic. Upon this hypothesis, which is a very useful one to work with, these authors separated all igneous and volcanic rocks into two classes, viz., the Basic and the Acid; the former containing from 45-58 per cent., the latter 62-78 per cent. of that mineral. But there are a few intermediate varieties which serve to bridge over the space between the Basic and Acid Groups. The following is a generalised arrangement of the most important rocks under the above heads:--

Tabular View of Chief Igneous and Volcanic Rocks.

BASIC GROUP.

1. Basalt and Dolerite. 2. Gabbro. 3. Diorite. 4. Diabase and Melaphyre. 5. Porphyrite.

INTERMEDIATE GROUP.

6. Syenite. 7. Mica-trap, or Lampophyre. 8. Andesite.

ACID GROUP.

9. Trachyte, Domite, and Phonolite. 10. Rhyolite and Obsidian. 11.

Granophyre. 12. Granite.

In the above grouping, and in the following definitions, I have not been able to follow any special authority. But the most serviceable text-books are those of Mr. Frank Rutley, Study of Rocks, and Dr. Hatch, Petrology; also H. Rosenbusch, Mikroskopische Physiographie der Mineralien, and F. Zirkel's Untersuchungen 對 er mikroskopische Structur der Basaltgesteine. We shall consider these in the order above indicated:--

1. BASALT.--The most extensively distributed of all volcanic rocks. It is a dense, dark rock of high specific gravity (2.4-2.8), consisting of plagioclase felspar (Labradorite or anorthite), augite, and titano-ferrite (titaniferous magnetite). Olivine is often present; and when abundant the rock is called "olivine-basalt." In the older rocks, basalt has often undergone decomposition into melaphyre; and amongst the metamorphic rocks it has been changed into diorite or hornblende rock; the augite having been converted into hornblende.

When leucite or nepheline replaces plagioclase, the rock becomes a leucite-basalt,[1] or nepheline-basalt. Some basalts have a glass paste, or "ground-mass," in which the minerals are enclosed.

The lava of Vesuvius may be regarded as a variety of basalt in which leucite replaces plagioclase, although this latter mineral is also present. Zirkel calls it "Sanidin-leucitgestein," as both the macroscopic and microscopic structure reveal the presence of leucite, sanidine, plagioclase, nephiline, augite, mica, olivine, apatite, and magnetite.[2]

Dolerite does not differ essentially from basalt in composition or structure, but is a largely crystalline-granular variety, occurring more abundantly than basalt amongst the more ancient rocks, and the different minerals are distinctly visible to the naked eye.

A remarkable variety of this rock occurs at Slieve Gullion in Ireland, in which mica is so abundant as to constitute the rock a "micaceous dolerite."

2. GABBRO.--A rather wide group of volcanic rocks with variable composition. Essentially it is a crystalline-granular compound of plagioclase,

generally Labradorite and diallage. Sometimes the pyroxenic mineral becomes hypersthene, giving rise to hypersthene-gabbro; or when hornblende is present, to hornblende-gabbro; when olivine, to olivine-gabbro. Magnetite is always present.

These rocks occur in the Carlingford district in Ireland, in the Lizard district of Cornwall, the Inner Hebrides (Mull, Skye, etc.) of Scotland, and in Saxony.

3. DIORITE.--A crystalline-granular compound of plagioclase and hornblende with magnetite. When quartz is present it becomes (according to the usual British acceptation) a syenite; but this view is gradually giving place to the German definition of syenite, which is a compound of orthoclase and hornblende; and it may be better to denominate the variety as quartz-diorite. The diorites are abundant as sheets and dykes amongst the older paleozoic and metamorphic rocks, and are sometimes exceedingly rich in magnetite. Mica, epidote, and chlorite are also present as accessories.

The rock occurs in North Wales, Charnwood Forest, Wicklow, Galway, and Donegal, and the Highlands of Scotland. There can be little doubt that amongst the metamorphic rocks of Galway, Mayo, and Donegal the great beds of (often columnar) diorite were originally augitic lavas, which have since undergone transformation.

4. DIABASE.--It is very doubtful if "Diabase" ought to be regarded as a distinct species of igneous rock, as it seems to be simply an altered variety of basalt or dolerite, in which chlorite, a secondary alteration-product, has been developed by the decomposition of the pyroxene or olivine of the original rock. It is a convenient name for use in the field when doubt occurs as to the real nature of an igneous rock. Melaphyre is a name given to the very dark varieties of altered augitic lavas, rich in magnetite and chlorite.

5. PORPHYRITE (or quartzless porphyry).--A basic variety of felstone-porphyry, consisting of a felspathic base with distinct crystals of felspar, with which there may be others of hornblende, mica, or augite. The colour is generally red or purple, and it weathers into red clay, in contrast to the highly acid or silicated felsites which weather into whitish sand.

6. SYENITE.--As stated above, this name has been variously applied. Its

derivation is from Syene (Assouan) in Egypt, and the granitic rocks of that district were called "syenites," under the supposition (now known to be erroneous) that they differ from ordinary granites in that they were supposed to be composed of quartz, felspar, and hornblende, instead of quartz, felspar, and mica. From this it arose that syenite was regarded as a variety of granite in which the mica is replaced by hornblende, and this has generally been the British view of the question. But the German definition is applied to an entirely different rock, belonging to the felstone family; and according to this classification syenite consists of a crystalline-granular compound of orthoclase and hornblende, in which quartz may or may not be present. From this it will be seen that, according to Zirkel, syenite is essentially distinct from diorite in the species of its felspar.[3] It seems desirable to adopt the German view; and as regards diorites containing quartz as an accessory, to apply to them the name of quartz-diorite, as stated above, the name syenite as used by British geologists having arisen from a misconception.

7. MICA-TRAP (LAMPOPHYRE).--A rock, allied to the felstone family, in which mica is an abundant and essential constituent, thus consisting of plagioclase and mica, with a little magnetite. Quartz may be an accessory. This rock occurs amongst the Lower Silurian strata of Ireland, Cumberland, and the South of Scotland; it is not volcanic in the ordinary acceptation of that term. The term lampophyre was introduced by Gabel in describing the mica-traps of Fichtelgebirge.

8. ANDESITE.--This is a dark-coloured, compact or vesicular, semi-vitreous group of volcanic rocks, composed essentially of a glassy plagioclase felspar, and a ferro-magnesian constituent enclosed in a glassy base. According to the nature of the ferro-magnesian constituent, the group may be divided into hornblende-andesite, biotite-andesite, and augite-andesite. Quartz is sometimes present, and when this mineral becomes an essential it gives rise to a variety called quartz-andesite or dacite.

These rocks are the principal constituents of the lavas of the Andes, and the name was first applied to them by Leopold von Buch; but their representatives also occur in the British Isles, Germany, and elsewhere. Dacite is the lava of Krakatoa and some of the volcanoes of Japan.

9, 10. TRACHYTE and DOMITE, etc.--These names include very numerous varieties of highly silicated volcanic rock, and in their general form consist of a white felsitic paste with distinct crystals of sanidine, together with plagioclase, augite, biotite, hornblende, and accessories. When crystalline grains or blebs of quartz occur, we have a quartz-trachyte; when tridymite is abundant, as in the trachyte of Co. Antrim, we have "tridymite-trachyte."

The trachytes occupy a position between the pitchstone lavas on the one hand, and the andesites and granophyres on the other.

(b.) Domite is the name applied to the trachytic rocks of the Auvergne district and the Puy de De particularly. They do not contain free quartz, though they are highly acid rocks, containing sometimes as much as 68 per cent. of silica.

(c.) Phonolite (Clinkstone) is a trachytic rock, composed essentially of sanidine, nepheline, and augite or hornblende. It is usually of a greenish colour, hard and compact, so as to ring under the hammer; hence the name. The Wolf Rock is composed of phonolite, and it occurs largely in Auvergne.

(d.) Rhyolites are closely connected with the quartz-trachytes, but present a marked fluidal, spherulitic, or perlitic structure. They consist of a trachytic ground-mass in which grains or crystals of quartz and sanidine, with other accessory minerals, are imbedded. They occur amongst the volcanic rocks of the British Isles, Hungary, and the Lipari Islands, from which the name Liparite has been derived.

(e.) Obsidian (Pitchstone).--This is a vitreous, highly acid rock, which has become a volcanic glass in consequence of rapid cooling, distinct minerals not having had time to form. It has a conchoidal fracture, various shades of colour from grey to black; and under the microscope is seen to contain crystallites or microliths, often beautifully arranged in stellate or feathery groups. Spherulitic structure is not infrequent; and occasionally a few crystals of sanidine, augite, or hornblende are to be seen imbedded in the glassy ground-mass. The rock occurs in dykes and veins in the Western Isles of Scotland, in Antrim, and on the borders of the Mourne Mountains, near Newry, in Ireland.

11. GRANOPHYRE.--This term, according to Geikie, embraces the greater

portion of the acid volcanic rocks of the Inner Hebrides. They are closely allied to the quartz-porphyries, and vary in texture from a fine felsitic or crystalline-granular quartz-porphyry, in the ground-mass of which porphyritic turbid felspar and quartz may generally be detected, to a granitoid rock of medium grain, in which the component dull felspar and clear quartz can be readily distinguished by the naked eye. Throughout all the varieties of texture there is a strong tendency to the development of minute irregularly-shaped cavities, inside of which quartz or felspar has crystallised out--a feature characteristic of the granites of Arran and of the Mourne Mountains.

12. GRANITE.--A true granite consists of a crystalline-granular rock consisting of quartz, felspar (orthoclase), and mica; the quartz is the paste or ground-mass in which the felspar and mica crystals are enclosed. This is the essential distinction between a granite and a quartz-porphyry or a granophyre. Owing to the presence of highly-heated steam under pressure in the body of the mass when in a molten condition, the quartz has been the last of the minerals to crystallise out, and hence does not itself occur with the crystalline form.

True granite is not a volcanic rock, and its representatives amongst volcanic ejecta are to be found in the granophyres, quartz-porphyries, felsites, trachytes, and rhyolites so abundant in most volcanic countries, and to one or other of these the so-called granites of the Mourne Mountains, of Arran Island, and of Skye are to be referred. Granite is a rock which has been intruded in a molten condition amongst the deep-seated parts of the crust, and has consolidated under great pressure in presence of aqueous vapour and with extreme slowness, resulting in the formation of a rock which is largely crystalline-granular. Its presence at the surface is due to denudation of the masses by which it was originally overspread.

###